The Sense of Smell

Author

R. H. Wright, Ph.D.
Vancouver, Canada

CRC Press, Inc.
Boca Raton, Florida

Library of Congress Cataloging in Publication Data

Wright, Robert H. (Robert Hamilton), 1906-
The sense of smell.

Includes index.
1. Smell. I. Title.
QP458.W73 152.1′66 82-1203
ISBN 0-8493-5232-0 AACR2

This book represents information obtained from authentic and highly regarded sources. Reprinted material is quoted with permission, and sources are indicated. A wide variety of references are listed. Every reasonable effort has been made to give reliable data and information, but the author and the publisher cannot assume responsibility for the validity of all materials or for the consequences of their use.

Direct all inquiries to CRC Press, Inc., 2000 Corporate Blvd., N.W., Boca Raton, Florida, 33431.

International Standard Book Number 0-8493-5232-0

Library of Congress Card Number 80-1203
Printed in the United States

PREFACE

The study of smells and smelling is the most interdisciplinary of the sciences. It is not enough merely to know the chemical formula of an odorous material because what matters equally is the message its odor brings to the organism perceiving it. We must approach the subject from a great many directions if we are to apprehend its role as a channel of communication between the outside world and the living organism. The part played by smells in the activities of insects or fishes or mice is as much a part of the picture as are the purely human aspects of the matter. To ignore the olfactory responses of other creatures would be like trying to comprehend human genetics without taking into account Mendel's work with garden peas or the myriad studies of fruit flies.

Between the odorous molecules at one end and at the other end the significance which the organism reads into the message these molecules bring to it, there are such intermediate matters as the anatomy of the receptor organs, the process of stimulus-receptor interaction, and the transmission to the Central Nervous System of a pattern of nerve impulses which is in one-to-one relation with the special attributes of the stimulating molecules. Finally, there remains the almost wholly unexplored matter of the way the incoming signal is compared with others stored in the memory or, perhaps, genetically fixed, so that a meaning can be assigned and the appropriate action taken.

In the past 25 years we have seen an outburst of symposia on the sense of smell whose published proceedings can be likened to boxes filled with the bits of a jig saw puzzle. The individual pieces have interesting shapes and patterns, but the whole picture cannot be more than guessed at unless the pieces are fitted together and the puzzle assembled.

That is what this book will try to do.

In Part I, I will present what I believe to be a theoretical synthesis which incorporates a greater range and variety of evidences than any other account so far offered. Some of the pieces of the puzzle are still missing, some are no doubt put in upside down, and some may be in the wrong place, but I believe the overall pattern can now be discerned and so the way may be made easier for those who must correct and complete the picture.

In presenting this picture, it is necessary to introduce facts and concepts from such diverse fields as chemistry and telephone engineering, spectroscopy and applied entomology, as well as anatomy, neurophysiology, and psychophysics, and so it is necessary to avoid as far as possible the specialized language of any one discipline. To this end, I have tried to emulate such masters as Faraday and Darwin whose command of the richness of ordinary English is infinitely preferable to that of those present-day writers by whom prolixity is equated with scholarship and obscurity with profundity.

In Part II, I will try to show the bearing of the synthesis achieved in Part I on the practical control of insect pests, or the practice of medicine, or the effective deployment of police dogs, as well as the control of air pollution or the psychological responses of humans to odors and the part played by odors in the lives of fish. Any one of these topics could be expanded into an entire book, but my object is not so much to exhaust each topic as to attract newcomers into a field that is unrivalled for all-round interest and the good company of intelligent people in just about every walk of life.

Finally, in Part III, which might be called "Unfinished Business", I mention some avenues still unexplored: "Things I would do were I thirty years younger and thirty times richer".

My book is evidently not the last word on the subject, nor is it the first. But I believe it to be the first to look at the subject from so many directions.

R. H. W.
Vancouver,
June, 1982

THE AUTHOR

Robert H. Wright received his undergraduate training in Chemistry and Physics at the University of British Columbia. He received his M.Sc. in Physical Chemistry and Physics in 1930, and his Ph.D. in Physical Chemistry in 1931, both from McGill University. In 1973 he received the D.Sc. (Hon. Causa), from the University of New Brunswick. He received the R. S. Jane Memorial Lecture Award presented by the Canadian Society for Chemical Engineering in 1976.

Dr. Wright was formerly head of the Olfactory Response Investigation at the British Columbia Research Council. He is the author of *The Science of Smell*, published in 1964, as well as numerous technical papers.

ACKNOWLEDGMENTS

I am profoundly grateful to the organizations whose financial support enabled me to proceed with the investgations which have made this book possible.

The British Columbia Research Council which provided the basic laboratory facilities and some part of the financial overhead has a central role in the story. The Canada Department of Agriculture through its extramural grants made it possible for the work to begin and the Canada Department of Forestry later joined to expand the range of matters included in it. Abbott Laboratories, Inc., the Trans-Mountain Oil Pipeline Company Ltd., and especially the Donner Canadian Foundation made it possible to work down into the more fundamental and theoretical aspects. At the same time, the U. S. Department of National Health and Welfare and the U. S. Army Medical Research and Development Command as well as the Canadian International Development Agency and the Food and Agriculture Organization of the United Nations provided funds for the exploration of practical applications without which the strictly scientific aspects would have been largely ornamental. Still more recently, an industrial firm which prefers to remain in the background has provided most welcome encouragement and support at a time when my forced retirement made it seem that my days of active research were forever gone.

And most of all, I thank my colleagues and those critics — both friendly and hostile — without whose input little would have been accomplished and who it would be difficult to name individually in view of the varied nature of their contributions.

. . . I have regarded the approbation of my confreres as nothing more than encouragement to do better; my studies have become more profound, I have meditated on the subject, and, in short, I have endeavored to change a mere dissertation into a monograph . . .

Hippolyte Cloquet, M. D. in the preface to
Osphresiologie ou Traite des Odeurs du Sens
et des Organes de l'Olfaction. Paris, 1821.

TABLE OF CONTENTS

PART I: PRINCIPLES

PART I
Principles

I have rather been desirous of discovering new facts and relations . . . than in exalting the force of those already obtained, being assured that the latter would find their full development hereafter.

M. Faraday

Chapter 1

A SCIENCE FOR ALL SEASONS

In a treatise on theoretical chemistry, widely different chapters of physics and chemistry must, of course, find their place; it must contain, in fact, all that the physicist must know of chemistry, and all that the chemist must know of physics, unless both are content to be specialists in their own science.

W. Nernst

Smelling is a complex process. It starts with molecules in the outside world which possess certain special attributes. By virtue of these attributes, the molecules can interact with sensory receptors in the nose so as to send specific nerve impulse-patterns to the brain. The brain then compares the incoming signal with other signal-patterns held in a memory store and, depending upon whether or not it can find a matching pattern, it assigns a meaning to the signal and decides on the most appropriate response.

If we are to understand the overall process, we have to begin by recognizing the various stages and use the appropriate techniques to investigate each one. We have to know structural-organic chemistry to characterize the stimulating molecules. We have to use physics as well as physical-organic chemistry to search for the special molecular attributes from which the special smell originates. The interaction process between the molecules of the stimulus and those in the sensitive nerve endings will eventually be made plain by, most probably, a mix of biochemistry and quantum physics. The signals passing from the primary receptors to the processing centers in the brain travel along nerves in the form of action potentials which can be recorded by techniques of electrophysiology. The mental processing mechanisms are a *terra incognita* into which the most promising lines of approach appear to be various branches of psychology combined with hints drawn from the world of computer design and programming. The physiological, psychological, and behavioral responses to the information brought to the organism via its nose are matters for both the behaviorists and the physiologists inasmuch as emotions of love or fear or fury can give rise to bodily changes and chemical emissions in the form of molecules with the ability to convey new messages — thereby completing the circle.

Any approach to the science of smells and smelling — sometimes called osphresiology — is bound to fail if it concentrates on only one or two of these aspects and ignores or dismisses the rest as unimportant. At the same time, it is difficult for anyone to develop real expertise in as many as two of these scientific disciplines. Under the circumstances, the important thing is for each specialist to make himself aware of and to comprehend the demands of each of the others. He must be ready to see that as far as possible his ideas are consistent with the established *phenomena,* though he need not always make them conform with the established *ideas* of the other disciplines.

The tendency to package science into disparate disciplines makes it difficult to approach osphresiology with the necessary broadness of outlook.

Before starting to gather the material for this book, I asked an entomologist what he knew of information theory and was met with a blank stare. A physicist knew little and cared less about the signals which regulate the behavior of insects. A physiologist dimly remembered hearing something about a quantum theory in his undergraduate days. A spectroscopist had never even conversed with a neurophysiologist, and a psychologist and a structural-organic chemist could not talk shop even with the help of an interpreter! Inasmuch as the sense of smell must draw on all these disciplines for its full understanding, this state of sensory deprivation is as regrettable as it is understandable. More important, it has had the unfortunate result of too many independent studies by specialists, each of whom has tried to present a comprehensive account based on his own special and therefore limited viewpoint. Without some insight into the experimental techniques and mental processes and constructs of the others, each specialist is walled off from fruitful exchanges with other kinds of specialist.

My task in writing this book is to open windows or at least to create peepholes in the walls in the hope that a process of cross-ventilation will bring a little fresh air into the rather stuffy sanctums of the specialists and, with a change of metaphor, bring about a cross-fertilization of ideas.

Of necessity, in pursuing this goal I will have to write for the psychologist about organic chemistry at a level and in a way that will make the chemist classify the book as something suitable only for the junior high school; and this will be true for each specialist when reading the parts dealing with his own field. The alternative is to write for each specialist in his own jargon; and this simply will not do.

Apart from the problem of language, perhaps the strangest feature of the various supposedly "interdisciplinary" symposia that have been published in recent years is the almost total absence of any reference to organic evolution. The cement needed to bring the various disciplines together must be the recognition of the evolutionary origin and probable development of chemoreception which is the most primitive of all the senses, and of olfaction which is at once the most sensitive, the most discriminating, and the most refined of the chemical senses.

A primitive, one-celled organism is clearly incapable of employing anything in the nature of a visual or auditory sense and must rely on purely tactile interactions with its environment. These can be either mechanical or chemical. By the time life had reached the level of the single cell there must have been a membrane with a structure intricate enough to allow various special functions (imbibition, excretion, enzyme-controlled conversions, etc.) to proceed at discrete sites distributed over the surface of the cell. Chemicals in the environment would then be able to act on the membrane in more than one way.

At the simplest level there would be a quasi-mechanical interaction arising from a combination of osmosis and altered surface tension. This would present a very generalized and undiscriminating insult to the organism, of the sort that continues to function in the higher vertebrates who feel a single, undifferentiated pain response when salt or alcohol or almost any water-soluble or irritant chemical enters a wound.

At a more specifically selective level, certain substances might interact with specific enzymes to alter the chemical balance at selected places in the membrane. Because enzymatic processes are usually stereospecific, interactions of this type tend to be selective as between enantiomeric (mirror image) stimuli. In the course of evolution, this type of interaction has probably culminated in the sense of taste which is known to display all-or-nothing responses to some enantiomeric stimuli.

These two types of interaction may be regarded as physical and chemical, respectively. What remains is the possibility of other interactions involving quantum effects. Such interactions would have been regarded as purely speculative a few years ago, but the development of the laser (to cite only one example) has shown that in a nonequilibrium system there can be large discharges of energy triggered by what would once have been regarded as quite inadequate stimuli. Living cells from their very nature are not in a state of equilibrium with their surroundings: at the very most there may be metastable islands in an ongoing sea of dynamism. It seems likely that quantum-induced processes in a cell membrane would be extremely local — more so than those due to mechanical deformations or altered enzymatic processes. The energy exchange in a quantum process would usually involve a single atom or interatomic bond, and an interaction whose primary effect is the rupture of a single chemical bond in a specific locality could well have great biological significance.

The living organism if it is to survive and reproduce must regulate its behavior in response to information brought to it in the form of signals from the environment. The more primitive the organism, the more it must depend upon chemical signals arriving through the kinds of interaction just described. By recognizing that chemical molecules can carry information to an organism and that the organism possesses the means to receive and act on the information, we establish the common ground on which the various specialists can meet and, given the will, communicate.

Chapter 2

ODORS AND INFORMATION

I often say that if you can measure that of which you speak, and express it by a number, you know something of your subject; but if you cannot measure it, your knowledge is meagre and unsatisfactory.

Lord Kelvin

Of the many scientific disciplines which help us to understand the sense of smell, the most unlikely candidate would seem to be telephone engineering or, more precisely, information theory.

In Figure 1 we have 16 messages represented by Xs. I ask you to pick out one to send to me and I then try to identify the correct one by asking questions which you can answer by doing something or not doing it. You can speak or not speak, move or not move, ring a bell or not ring it. You do not have to say "No", because not saying "Yes" conveys the same information. I could go through the array of Xs one by one, asking, "Is it this?" until I happen on the right one. With good luck I might get the message at the first try and with bad luck not until the end, so that on the average it would take about 8 questions to isolate any one of the 16 possibilities.

By proceeding systematically, I can infallibly eliminate all but the correct message with exactly four questions. To do this I draw a line through the middle of the array and I ask, "Is it on this side of the line?". Your response or lack of response immediately reduces my uncertainty by half. I next draw a line to divide the eight possibilities that remain and repeat the question. In this way, four properly chosen yes-or-not-yes bits of information will assuredly enable me to identify any one of 16 possible messages. Had there been 32 Xs to begin with I would need five questions instead of four, which is a trifling increase which becomes even more trifling as the number of Xs increases. Assuming that there are no errors in the transmission and reception of the signals, 20 yes-or-not-yes signals can specify more than a million messages.

Let us see what this has to do with smells and smelling.

Among the higher vertebrates, dogs and some especially gifted or trained humans can achieve remarkable feats of discrimination using their noses. For example, Professor Hans Kalmus[5] used London police dogs to show that every human being apparently has a body odor that is as uniquely personal as his fingerprints. How many bits of information would it take to identify any one human being out of the whole population of the earth?

Assuming that there are just as many men and boys in the world as there are women and girls, we could begin by asking, "Is the person female?". If there is no response we know that our target individual is male and have eliminated half the possibilities. For our next question, we will suppose that we know that half the male population is 19 years old or older, and so we ask, "Is he under 19?". If the answer is "Yes" we have once more halved the field. With just two questions we have eliminated from consideration three quarters of the population of the globe! Let us suppose that the population of the world is just over a billion (10^9); 30 properly framed questions will eliminate all but one because if you halve a billion (or, more precisely, 1,073,741,824); 30 times, the answer is 1. In technical terms, therefore, it will take *not less* than 30 bits of information to identify any one person out of a billion equally probable possibilities.

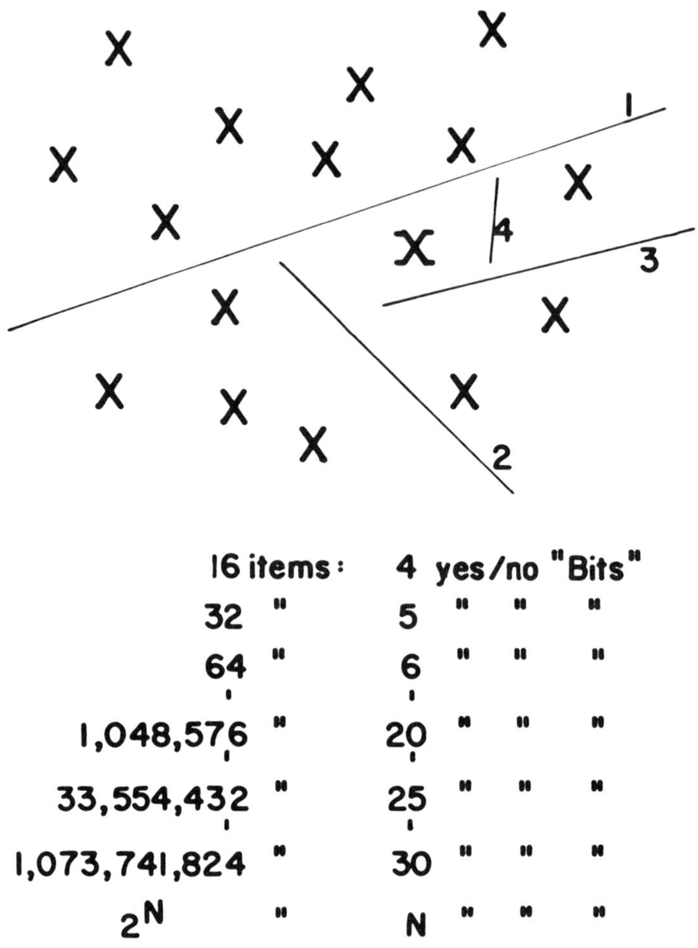

FIGURE 1. The amount of information in a message is the number of bits needed to distinguish that message from other equally probable messages. The fact that the nose can discriminate between many thousands or millions of smell sensations requires an informational input of at least 20 bits.

It may be objected that Professor Kalmus' experiments with no more than a dozen or so people do not prove that two individuals *cannot* present precisely the same body scent to a dog, although the use of several pairs of identical twins in the experiments makes it very probable.

Let us therefore consider the corresponding problem at the human level.

Professional perfumers are trained to evaluate the odorous properties of carefully purified chemicals, and in the course of their work they may have occasion to smell many thousands of such chemicals. More than one perfumer has stated that *in his experience* no two chemicals are precisely alike. This does not exclude the possibility that different chemicals may present identical odors, but it renders it highly improbable. The situation is parallel with that of the fingerprint expert who can say that *in his experience* no two fingerprints are precisely alike. To make that a categorically affirmative assertion he would have to examine the print of every finger that ever was — which is clearly impossible. The smell expert and the fingerprint expert therefore have to fall back on the laws of probability. It then turns out that if we were to examine the odors of 10,000 chemicals or the prints of 10,000 fingers without finding two that

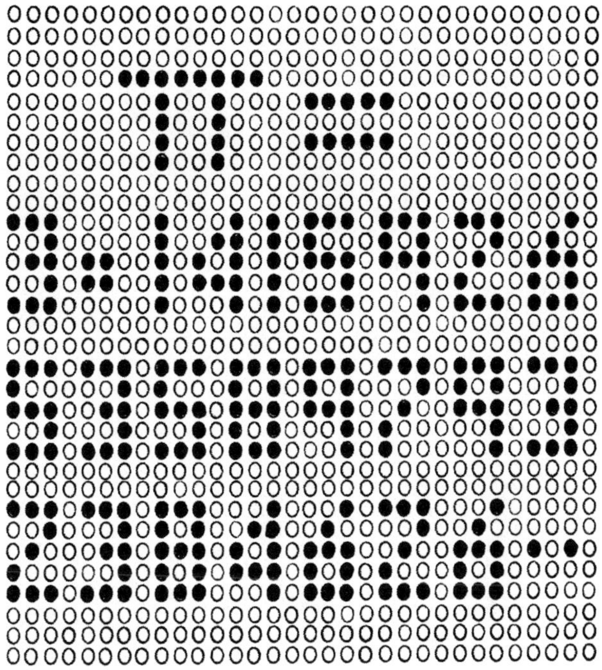

FIGURE 2. A 32×32 array of circles which can be black or not-black can present 1024 bits of information and convey 2^{1024} or 1.8 $\times 10^{308}$ different messages. Mental processing may interpret this display as the value of pi to 22 places of decimals.

were indistinguishable, we would probably have to look at 10,000 times 10,000 to have an even chance of finding such a pair. The calculation is as follows:

Consider a population, M, made up of N equally probable types. Let a sample, S, be drawn from M, S being much smaller than M. The probability, P, that no two of S are the same is

$$P = \frac{N-1}{N} \times \frac{N-2}{N} \times \frac{N-3}{N} \times \ldots \times \frac{N-S+1}{N}$$

$$= \frac{N!}{N^S(N-S)!}$$

for S equal to or less than N.

If S and N are both large

$$P = \exp\left(\frac{-S^2}{2N}\right)$$

so that if S = 10,000 with no two alike

$$N \approx S^2$$

There is no way that a sensation can be *more specific* than the stimulus which produces it, and from this it necessarily follows that the nose can make very precise discriminations between many millions of smell sensations.

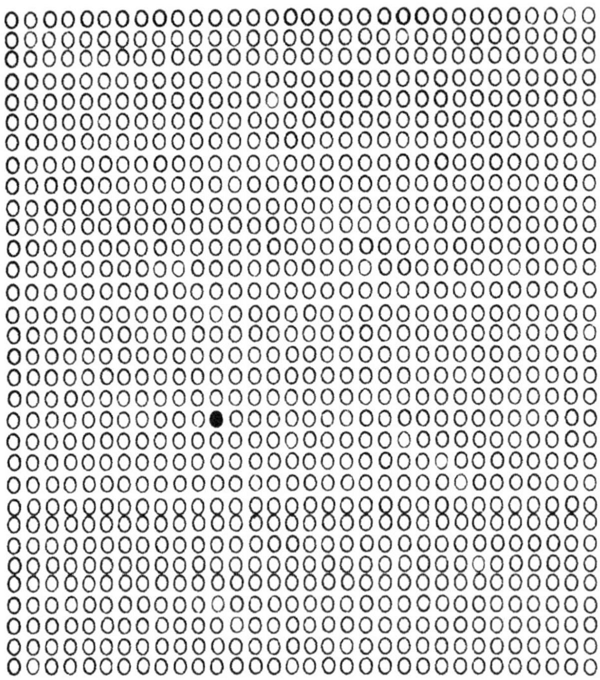

FIGURE 3. This display conveys exactly the same number of bits of information as Figure 2. If it seems less interesting it is because it is less evocative of interesting associations.

Scientifically, it is always preferable to express our ideas in quantitative and mathematically exact language, if only because mathematics has no symbols for confused ideas. By applying the basic principles of information theory the words "very precise" can be given a very precise meaning: the quantity of "information" in a signal (of whatever sort) is specified by a specific number of bits.

Note, however, that the use an organism makes of the information takes us into the most rarefied realms of psychology.

Let me illustrate this with an array of 1024 circles which, for convenience, can be set out in a 32 × 32 square. Since each circle can be black or not-black, the array presents 1024 bits of information and can transmit 2^{1024} messages since it allows the immense number of 1.8×10^{308} different patterns. Figure 2 presents one of these patterns. To a mathematician who counts in tens and uses arabic numerals it tells him the value of pi to several places of decimals. The next (Figure 3) presents exactly the same amount of information as the first, but to most people it will be much less interesting, though it would be immensely important if it indicated the position of a buried treasure.

Evidently, information as understood by the telephone engineer means something different from information as understood by the reader of an encyclopedia. In the first case it means, simply and only, a means of specifying which one of several alternative possibilities has to be transmitted *and that is all.* In the second case, as soon as a particular pattern is specified, the brain collates it with other learned or remembered patterns and gives it a meaning. A person who counts in tens and uses arabic numerals will interpret Figure 2 as signifying the ratio of the circumference to the diameter of a circle. A visitor from another plant with 12 fingers and a different set of written symbols would find that display meaningless, but would instantly recognize one of the other 10^{308} patterns which presented pi in a manner familiar to him.

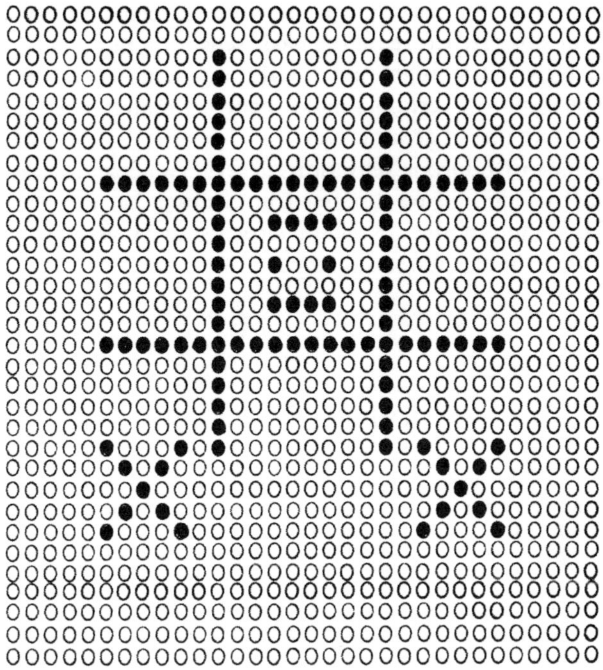

FIGURE 4. A display pattern can mean different things to different people, depending upon their past experience. This might suggest a game of tic-tac-toe, a football game, or the Chinese character for "his, hers, its".

The ability of the individual to assign his own meaning to a signal is illustrated in Figure 4. Some people would think it was a game of tic-tac-toe. To others it would suggest a football play. To still others it could suggest the Chinese character 其.

The point of applying information theory to the phenomena of smelling is to show that if the olfactory receptor system can transmit a sensory input made up of some 30 yes-or-not-yes elements, there must be some 30 molecular attributes whose presence or not-presence determines the specific input from each kind of molecule. This conclusion, which seems inescapable, has been rather strangely disregarded by many investigators. What it means in practice is that we can rule out the kind of approach which tries to associate odorous character with such molecular attributes as oxydizing potential, dipole moment, molecular weight, or vapor pressure. The molecular attributes we have to look for must form patterns in which are embodied at least 30 elements whose presence or absence will *both* be significant in determining the overall pattern.

The pattern is complex because of the number of elements that go to make it up. The elements themselves, taken singly, can be relatively simple and, indeed, must *be* simple when we remember their evolutionary history. What is important is the requirement that there should be a minimum of overlap between elements so that as far as possible the receptor apparatus can determine whether an element is there or not-there.

Information theory cannot tell us what those elements are, but it can tell us what kind of thing they are not.

Some psychologists will probably object that the proper way to measure the capacity of the human nose to transmit information is to equate it with the ability to name the stimulus correctly. Using this criterion, the capacity works out at somewhere between 4 and 6 bits. Bearing in mind that the brain has first to register a pattern, then compare it with patterns stored in the memory, and then select a name for the sensation by searching the vocabulary for the appropriate word, it seems clear that this is superimposing an irrelevant superstructure on the basic sensory input. A resident of the tropics or of downtown New York might not be able to name correctly the odor of pine needles, but he could almost certainly make a correct match in a triangle test with two jars of pine needles and one of cedar leaves.

Verbal descriptions of odorous sensations can vary in curious ways, a phenomenon called "parosmia" which will be discussed in Chapter 24. For the present, it seems appropriate to limit the amount of mental processing by asking simply, "Is this the same as that?" rather than, "What is this?".

REFERENCES

1. **Benarie, M. M.,** Language as a model relating to odor quality, *Atmos. Environ.,* 7, 369, 1973.
2. **Desor, J. A. and Beauchamp, G. K.,** The human capacity to transmit olfactory information, *Percept. Psychophys.,* 16, 551, 1974.
3. **Hainer, R. M., Emslie, A. G., and Jacobson, A.,** An information theory of olfaction, *Ann. N.Y. Acad. Sci.,* 58, Art. 2, 158, 1954.
4. **Jones, F. N.,** The informational content of olfactory quality, in *Theories of Odor and Odor Measurement,* Tanyolac, N. N., Ed., Robert College Research Center, Bebek, Istanbul, 1968, 133.
5. **Kalmus, H.,** The discrimination by the nose of the dog of individual human odors and in particular of the odors of twins, *Br. J. Anim. Behav.,* 3, 25, 1955.
6. **Pierce, J. R.,** *Symbols, Signals and Noise, The Nature and Process of Communication,* Harper & Row, New York, 1961.

Chapter 3

ATTRIBUTE STATISTICS

A connection frequently exists between physical and chemical properties, and in many cases recourse must be had to both for an explanation of a phenomenon to which they both contribute.

C. L. Berthelot

A substance such as muscone (a macrocyclic ketone) smells like things that are chemically and stereochemically very different, such as, for example, a benzene ring with three nitro groups, a methyl group, and a tertiary butyl group hanging from it as illustrated in Figure 1. The chemical structure as embodied in the structural formula necessarily determines *all* the properties of the molecule including the odor, but it is clear from the example of the musks and many others that the particular atoms and chemical linkages making up the structure are not what the receptor system senses and registers as a specific smell sensation. Many years ago, G. M. Dyson[2] made a most thorough and detailed attempt to correlate the odors of compounds with their chemical structures as expressed in their structural formulas, or wiring diagrams. At the end he could only conclude that, "No chemical data, either from the viewpoint of reactivity or chemical structure will give us the key to the rational, quantitative interpretation of odor phenomena."[2] To this day, no one has seriously challenged that statement.

Lock-and-key mechanisms which probably underlie the phenomena of immunology have been suggested as perhaps accounting for the ability of the nose to register a specific response to each specific stimulus. But lock-and-key mechanisms suppose that the odorous molecules register their specific attribute-patterns by settling themselves into sites or sockets of the matching shape. Such a site-fitting process would seem to need as many kinds of locks as there are keys, which is hard to accept. What is even harder to accept is the implication that locks exist before there are keys to fit them. Many novel synthetic perfumes have only appeared in the world within the last 100 years yet their distinctive odors were immediately apparent to the chemists who made them. Evolutionary progress depends upon what has worked in the past rather than upon what may be required to work in a distant and unpredictable future.

Thus the pattern-elements must be molecular attributes which can turn up as readily in a synthetic novelty as in a naturally occurring molecule. What this means is that each odorous molecular species must be endowed with attributes which constitute a pattern, some elements of which may appear in other molecular species but which, taken together, form a distinctive pattern.

The real existence of such distinctive but at times overlapping attribute-patterns was first demonstrated through a study of the way certain insects are attracted to certain chemicals. For example, 2426 compounds were tested with the oriental fruit fly, *Dacus dorsalis* Hendel, and, in separate experiments, with the Mexican fruit fly, *Anastrepha ludens* Loew. Of the 2426 chemicals, 538 or 22.1% attracted the oriental fruit fly and 194 or 8.0% attracted the Mexican fruit fly. If we can regard the 538 oriental fruit fly attractants as a random subset of the 2426 compounds, then 8.0% of the 538, or 43 compounds in the subset should attract the Mexican fruit fly. The observed number was 44, which is statistically indistinguishable from the number predicted for the subset.

This can only mean that from the point of view of the Mexican fruit fly, the 538

FIGURE 1. The model at the left represents the arrangement of atoms in a molecule of muscone, the chemical secreted by the musk deer and representing the musk odor in its purest form. At the right is a model of toluene musk, a synthetic compound with little or no chemical resemblance to natural musk but with an odor which, though not identical with that of natural musk, is similar enough to be useful in certain branches of perfumery.

oriental fruit fly attractants really are a random subset of the 2426. Therefore the specific molecular attributes which affect one insect can exist in the molecules along with but independently of the attributes which affect a different species of insect. This is not an isolated example because similar results have been found with more than a dozen pairs of insect species as shown in Table 1.

What is especially significant is the fact that the same experimental technique has shown that while each species of insect responds to a certain specific pattern of attributes, the pattern for one insect may at times include some of the pattern-elements which are part of a different pattern specific to a different insect.

For example, 2017 compounds were tested as attractants for the oriental fruit fly and for the melon fly, *Dacus curcurbitae* Coq. Using the same method of analysis as before, 83 of the 2017 were expected to attract both species of insect but the experiments showed that there were in fact 172 doubly-effective chemicals. Evidently the two patterns share one or more elementary attributes so that the compounds which attract one insect are not a random subset of the total so far as the other insect is concerned. By virtue of the shared attribute(s), the fact that they attract one, means that they have a better-than-chance prospect of attracting the other.

This, again, is not an isolated example inasmuch as similar results have been obtained with at least five other species as shown in Table 2. The fact that in the pair, oriental fruit fly — boll weevil, the number of double-effectives is less than the number expected would appear to suggest that pattern-elements may at times be mutually exclusive. An alternative explanation will be presented later, in Chapter 7.

Accepting the fact that the required molecular attributes or pattern-elements are not mutually exclusive, we must conclude that a molecule either possesses attribute A or it does not, and this is independent of whether or not it possesses attributes B, C, and so on. If anything further were needed to eliminate lock-and-key mechanisms based on molecular shape, this would be nearly sufficient.

The qualification "nearly" is eliminated by one further fact.

The American scientists made tests with 1684 compounds for the ability to attract *three* species of insect: the oriental fruit fly, the melon fly, and the Mediterranean fruit fly, or medfly, *Ceratitis capitata,* Wied. There is overlap between each pair of attribute-patterns because the number of doubly-effective compounds was always larger than expectation. When we come to the triple attractants the number found was nearly five times the number expected. A site-fitting mechanism could account for double

Table 1
STATISTICS OF CHEMICALS WHICH ATTRACT DIFFERENT INSECT SPECIES BY VIRTUE OF INDEPENDENT MOLECULAR ATTRIBUTES

	Number of compounds		
		Attractants	
Insect pair	Tested	Found	Expected
Oriental fruit fly Mexican fruit fly	2426	44	43
Melon fly Mexican fruit fly	1509	27	25
Melon fly Boll weevil	177	3	4
Medfly Mexican fruit fly	2089	48	49
Medfly Boll weevil	347	13	14
Mexican fruit fly Boll weevil	349	4	4
Oriental fruit fly European chafer	118	7	7
Melon fly European chafer	61	1	2
Melon fly Pink bollworm	152	12	9
Medfly Pink bollworm	281	33	32
Mexican fruit fly European chafer	99	1	1
Mexican fruit fly Pink bollworm	445	5	4
European chafer Pink bollworm	90	8	8
European chafer Boll weevil	98	5	6
Pink bollworm Boll weevil	383	27	25

Table 2
STATISTICS OF CHEMICALS WHICH ATTRACT TWO SPECIES OF INSECT BY VIRTUE OF OVERLAPPING MOLECULAR ATTRIBUTES

	Number of compounds		
		Attractants	
Insect pair	Tested	Found	Expected
Oriental fruit fly Melon fly	2017	172	83
Oriental fruit fly Medfly	2421	227	165
Melon fly Medfly	1685	92	58
Oriental fruit fly Pink bollworm	476	33	24
Medfly European chafer	74	12	7
Oriental fruit fly Boll weevil	368	9	16

FIGURE 2. The difficulty with site-fitting theories of odor specificity! The can will fit well into two sites (box and pan), but not into the slot; the saucer will fit the pan and the slot, but not the box; and the book will fit the slot and the box, but not the pan. It would be difficult to contrive a shape that would fit snugly into all three sites, yet odorants can show precisely this kind of overlapping behavior.

Table 3
STATISTICS OF CHEMICALS WHICH ATTRACT THREE SPECIES OF INSECT THANKS TO OVERLAPPING MOLECULAR VARIATION PATTERNS

	Number of compounds		
		Attractants	
Insect	Tested	Found	Expected
Oriental fruit fly	1684	404	—
Melon fly	1684	229	—
Medfly	1684	427	
Oriental fruit fly Melon fly	1684	131	55[a]
Oriental fruit fly Medfly	1684	177	102
Melon fly Medfly	1684	91	58
Oriental fruit fly Melon fly Medfly	1684	68	14

[a] $\frac{404 \times 229}{1684} = 55$

attractants, but can scarcely be stretched to accomodate triple ones. In Figure 2 the tin will fit into the box and the pan, the saucer will fit into the pan and the slot, and the book will fit into the slot and the box. This is essentially the position as regards substances in the upper part of Table 3. But, again referring to Figure 2 the can will not fit the slot, the book will not fit the pan, and the saucer will not go snugly into the box. This is exactly contrary to the situation in the lower part of Table 3.

Thus the kind of molecular attribute which can serve as an olfactory pattern-element must be something comparable to the notes on a piano which represent more or less narrow bands of frequency in a linear continuum of frequencies. Anyone who has tuned a radio receiver knows that the various broadcasting stations are spread along the dial but each uses only a small part of the total spectrum of frequencies so as not to interfere with the others. A receiver tuned to one ignores the others. Conceivably, a traveller waking up in an unknown city could identify it without leaving his room by ascertaining what frequencies are used by the local broadcasting stations which will, in most cases, form a pattern which is peculiar to each locality. We might compare the receptor system in the nose to a number of receiving sets tuned to a series of frequency-bands. Each receiver contributes 1 bit of information either by receiving or by not-receiving a signal. As we have seen, 30 differently tuned receivers would be able to identify any one of a billion different cities or smells!

REFERENCES

1. **Amoore, J. E.,** *Molecular Basis of Odor,* Charles C Thomas, Springfield, Ill., 1970.
2. **Dyson, G. M.,** Odor and constitution among the mustard oils. VII. Reactivity and odor, *Perfum. Essent. Oil Rec.,* 22, 278, 1931.
3. **Wright, R. H.,** The investigation of primary odors, in *Theories of Odor and Odor Measurement,* Tanyolac, N. N., Ed., Robert College Research Center, Bebek, Istanbul, 1968, 459.
4. **Wright, R. H.,** Primary odors and insect attraction, *Can. Ent.,* 98, 1083, 1966.

Chapter 4

MOLECULAR VIBRATION

Advances in knowledge are not commonly made without the previous exercise of some boldness and license in guessing.

W. Whewell

The comparison just drawn between an array of separately tuned radio receivers and the sensitive sites in the nose may be more realistic than might appear at first sight, and in this and the following Chapters the theme will be developed in detail. Here, we will begin by looking more closely at the kind of physical attributes which seem best adapted to meeting the demands of information theory as set out in Chapter 2.

In Figure 1 two molecular models are compared, one is of the toluene musk already introduced in Figure 1, Chapter 3, and the other of the very similar musk xylol. At the left is the space-filling model which shows the overall shape as it would appear when viewed by its neighboring molecules. Superficially it resembles a handful of grapes without enough character to account for its odorous resemblance to a second and differently shaped handful.

If we could work our way into the structure so as to look at the molecule from the inside, we would arrive at the skeletal model shown at the right. It reveals musk xylol as having a flexible structure made up of mass points joined by elastic forces. Such a structure can be elastically deformed and will spring back and vibrate in as many ways as there are ways to deform it. Thanks to its springiness, the molecule has many modes of vibration, each with a specific frequency. The various modes can be set shaking separately or simultaneously, like the strings of a piano, and together they make up a pattern which is not likely to be matched in all its details by the pattern of any other molecule or musical chord. Individual frequencies, like individual notes on the piano, may contribute to many different patterns.

A structure as complex as that of musk xylol can evidently vibrate in very complex ways. Nevertheless, it can be shown that for a molecule made up of N atoms, the number of distinctive ways it can vibrate — what are called its Normal Modes — is given by the formula, $3N - 6$. (For the special case of a molecule whose atoms lie in a straight line, the number is $3N - 5$.) The molecule of musk xylol contains 36 atoms and so has 102 normal modes or the potential ability to display 102 frequencies to be tuned in by 102 different receiving sets. This is many more than the number of bits required to endow the compound with a unique attribute pattern, but it turns out that not all the vibrational modes can in fact be osmically active and contribute to the odor.

This is because the vibrations are quantized. That is, a normal mode of vibration requires a definite amount of energy to excite it, not less and also not more unless the excess amounts to 2 times or 3 times . . . the basic quantum. This basic quantum of energy, $E\nu$, depends upon the frequency (number of to-and-fro movements per second) of the oscillation, according to the formula

$$E_{\nu} = h\nu \quad \text{erg}$$

where h is Planck's constant and has a value of 6.5×10^{-27} erg × sec. Its dimensions are ergs (or energy) times time, so that when multiplied by a number (of oscillations) divided by time, the result will simply be energy.

Mathematics is a language, and translated into English this formula tells us that if

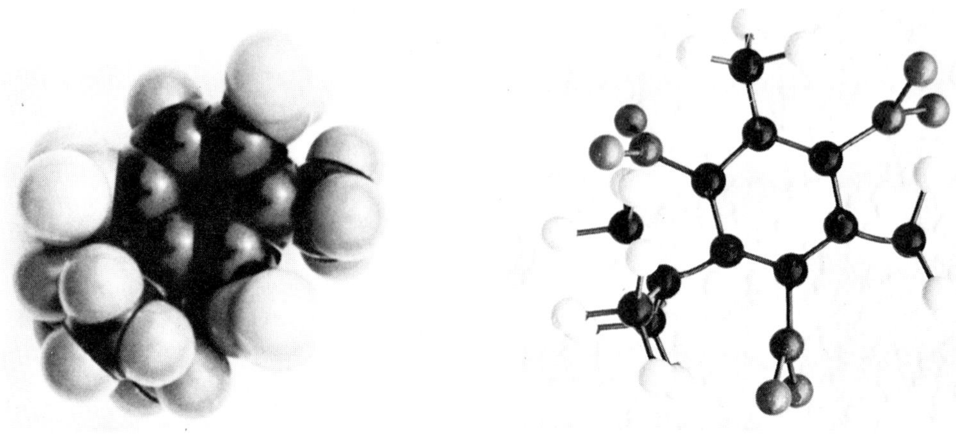

FIGURE 1. Two ways to look at a chemical molecule: the space-filling model at the left emphasizes its gross shape where the skeletal model at the right brings out its springiness. By virtue of the elasticity of the interatomic bonds, the structure can execute very complicated vibrational movements. These complex oscillations can, however, be resolved into a limited number of normal modes, each with a specific frequency of vibration. Together, these normal mode frequencies present a pattern which is unique for each species of molecule.

the frequency, ν, is large there is a lot of energy stored in the vibration and if it is small it takes much less energy to excite it.

Now, the energy to excite a vibration must come from somewhere, and in the world around us the only place a molecule can normally get the required energy is from collisions with other molecules. (There are a couple of other possible sources which I will consider shortly, but for the moment let us consider only collisional excitation.) We live in what is, cosmically speaking, a rather cold environment, about 25°C or, say, 300° absolute. Compare this with the surface of the sun which is at about 6000° absolute or the deep interior of the sun which is at many millions of degrees above absolute zero. Since the velocities of molecules depend upon their average temperature, at 300° absolute they are mostly moving rather slowly and so the energy they can put into a collision with an elastic structure is generally small. From this it follows that only the easily excited, low-frequency, low-energy vibrations of molecules are likely to be excited at the ordinary temperature.

Once again, let us look at the quantitative significance of this qualitative word, "likely".

At a temperature of 300° absolute (or any other stated temperature), molecules are not all moving with exactly the same speed. Some are moving quickly and some rather slowly, but the *average* velocity can be calculated precisely from the kinetic theory of gases. An unusually fast-moving molecule can, of course, excite a high-energy vibration in something it bumps into, but at 300° absolute there are not very many fast movers. Therefore only a few high-energy, high-frequency molecular oscillations are likely to be excited. The formula giving quantitative meaning to the word "likely" is a famous one discovered by Max Planck:

$$\left.\begin{array}{l}\text{Average number of}\\ \text{quanta in a normal}\\ \text{mode with a frequency, } \nu\end{array}\right\} = \frac{1}{e^{\frac{h\nu}{kT}} - 1}$$

In ordinary English, this formula says that at any temperature, T, the average num-

Table 1
AVERAGE NUMBER OF VIBRATIONAL QUANTA IN THE VARIOUS FREQUENCY MODES OF A MOLECULE AT 300 K

Frequency		
Hertz (to-and-fro movements/sec)	Wave number (Hertz divided by 3×10^{10})	Average number of vibrational quanta per molecule
30×10^{12}	1000	0.008
24×10^{12}	800	0.022
18×0^{12}	600	0.059
12×10^{12}	400	0.17
6×10^{12}	200	0.62
3×10^{12}	100	1.62
1.5×10^{12}	50	3.69

Note: The actual frequency in cycles per second, or Hertz, is an inconveniently large number. By dividing it by the velocity of light (3×10^{10} cm/sec) we arrive at a more easily handled number which is called the "wave number" because it represents, in fact, the number of waves per centimeter in the train of radiation with that frequency travelling with the speed of light. The word "frequency" is commonly used for both Hertz and wave number but there is little risk of confusion because the kind of frequency referred to is immediately evident from the size of the number.

ber of quanta in a normal mode with a frequency, ν, can be calculated, and the result of that calculation when T = 300° absolute is shown in Table 1.

From this table we see at once that there is very little likelihood of a normal mode having even one quantum of energy unless the frequency is less than about 800 wave numbers. Normal mode frequencies usually range from as low as 100 wave numbers up to about 4000 wave numbers, so that it is clear that not all the 93 normal mode vibrations of toluene musk, or of any other odorous molecule, will have to be considered. We have to direct our attention specifically to the low-frequency modes.

In the next chapter we will look at the ways the vibrational frequencies of molecules can be identified and mapped.

Chapter 5

MAPPING THE FREQUENCIES

If ever there was a flank movement on Nature by which she has been compelled to surrender a part of her secrets, it was the discovery of the spectroscope, which enables us to peer into the very heart of Nature.

R. C. Kedzie

In the last chapter, I explained that a molecule can become vibrationally activated as a result of a sufficiently violent collision with another molecule. I also mentioned that there are other ways for a molecule to become excited, and in this chapter one of these will be described because it provides the means whereby we can come to know the frequencies which an odorous molecule can present to the sensitive receptors in the nose.

Molecules enclosed in a vessel with opaque walls and at a uniform temperature can be set vibrating *only* by collisions. If, however, the walls are transparent and radiant energy can flow through from a hotter source, then the molecules may take energy from the passing stream in much the same way that a hydroelectric power dam can derive energy from the water moving past it. There is, nevertheless, an important difference between the stream of water and a stream of electromagnetic radiation. The difference represents a still-unexplained paradox of modern physics.

If you want to know where a beam of light or of any other kind of electromagnetic radiation will go, you picture it as a train of waves and you can then calculate precisely how it will behave after passing through a slit or bouncing off a mirror. If you want to know what the energy in the light ray will do when it interacts with the atoms or molecules of matter, you must picture the energy as moving in packets or quanta small enough to be swallowed whole by something as small as an electron! For practical purposes, this seeming contradiction is unimportant because where the radiation goes and what it does when it gets there are two entirely different matters.

The reason for describing these two aspects of radiation — its wave and its particle aspects — is that by regarding it as a wave motion we can go on to measure its wavelength. The speed of light is generally represented by the letter ''c'', nowadays familiar to nearly everyone from Einstein's formula

$$e = mc^2$$

The wavelength, usually represented by the Greek letter, λ, (lambda) can be measured in various ways which it is not necessary to describe here. Knowing the speed and the wavelength of light, the frequency, ν, or number of ups-and-downs of the wave passing a given point in 1 sec, can be calculated from the formula

$$c = \nu\lambda$$

In the last chapter, the term ''wave number'' was introduced. It is simply the reciprocal of the wavelength, or

$$\text{Wave number} = \frac{1}{\lambda} = \frac{\nu}{c}$$

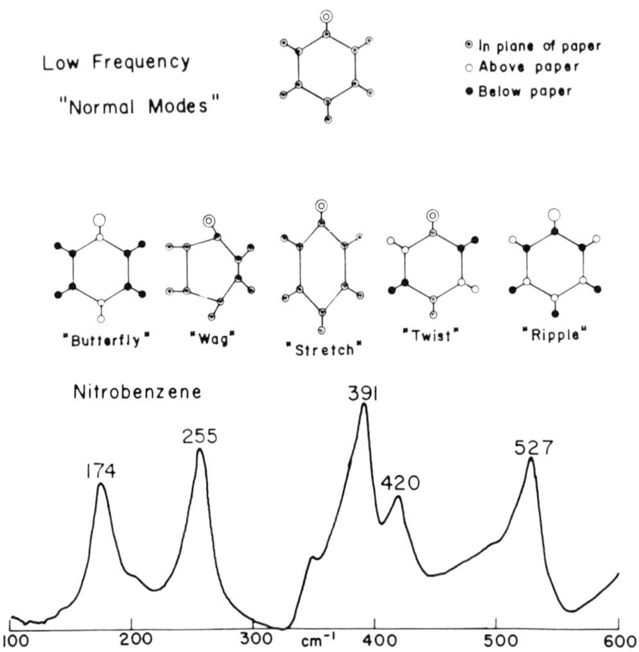

FIGURE 1. The far infrared absorption spectrum of nitrobenzene and the vibrational assignments or kinds of oscillation believed to produce each kind of absorption.

and is in many ways a more convenient quantity for some purposes than the true frequency because it is derived directly and simply from the measured wavelength, λ. It is called the wave number because it is the number of waves in 1 cm of the wave train.

This discussion has been leading up to the fact that, thanks to the wave aspect of radiation, we can assign a certain frequency to the energy radiated in any part of the spectrum. Hence, if we pass a stream of radiation containing many wave lengths or frequencies through a sample of matter and find that certain frequencies are removed from the passing stream, we know the size of the quanta of energy in which those particular molecules are interested. Once again, let us look for a more precise way to describe or specify that interest.

The parts of a vibrating molecule move to-and-fro a certain number of times per second, that is, with a certain frequency. The waves of radiation passing the molecule make a certain number of ups-and-downs per second. Should the frequency of these ups-and-downs in the radiation match the tos-and-fros needed to set the molecule vibrating, the molecule may take from the passing stream the energy carried by waves of that precise frequency. Hence, if the energy going in has a broad spectrum of wavelengths, what comes out will have been robbed of certain wavelengths by the molecules in its path, and the frequencies of the absorbed or missing radiation are the same as the vibrational frequencies that have been set in motion by the energy stolen from the stream of radiation.

A machine for measuring the absorption of radiation by chemical molecules is called an absorption spectrophotometer. It is not necessary to describe these machines in detail because all that is really needed at this point is to understand that by recording the wavelengths that are absorbed we can know the vibrational frequencies of the molecules doing the absorbing.

Yet it is necessary to make one further point. As explained in the last chapter, at the ordinary temperature and in the absence of an outside source of radiated energy, the only molecular vibrations likely to be important for our purposes are those of rather low frequency. The radiation that is absorbed at those frequencies falls in the far infrared part of the spectrum, and until recently spectrophotometers for work in the far infrared could not be bought commercially and each researcher had to design and build his own.* This was (and is) something that few people interested in osphresiology were willing or able to do. (How many bacteriologists can design and build a microscope?) Hence it is only since the early 1960s that the relevant vibrational attributes of odorous molecules could be critically examined.

Figure 1 shows the far infrared spectrum from 600 to 100 wave numbers of a fairly simple compound, nitrobenzene, which has the formula, $C_6H_5NO_2$, and the structure:

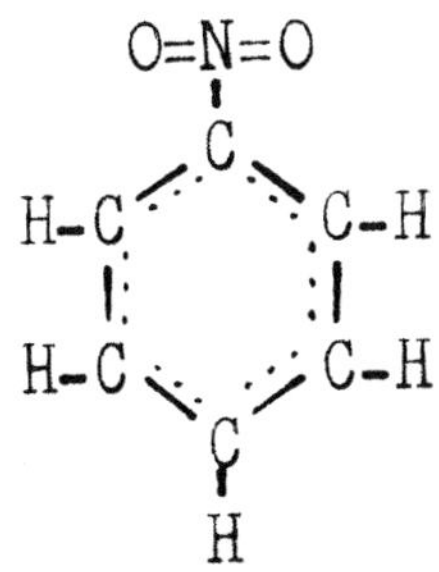

By virtue of its containing 14 atoms, nitrobenzene has 36 ($= 3 \times 14 - 6$) normal modes of vibration of which five have frequencies that appear in this part of the spectrum. The kinds of vibrational movement that give rise to each absorption band are believed to be those shown in Figure 1. Some are easier to visualize and describe in words than others. There are, naturally, modes of vibration which give rise to absorptions in other parts of the spectrum. To mention only one, there is the breathing mode with a frequency near 1000 wave numbers in which the ring expands and contracts.

It would be interesting to draw similar pictures of the normal modes of compounds as complex as the musks shown in Figure 1, Chapter 3, but the existing state of the art does not make this possible. Fortunately for the science of osphresiology, it is not necessary to specify the mechanical movement or make the vibrational assignments in detail in order to establish low-frequency molecular vibrations as the attributes which put bits of information into a smell signal. It is probable, however, that in the time to come a knowledge of the vibrational assignments will be needed if we are to understand the fine details of stimulus-receptor interaction.

Meanwhile, let us proceed to see how specific vibrational patterns can be correlated with the smell sensations an organism perceives.

* In the years before 1960 and going back to the 1930s, something called the Raman Effect could be used to chart the vibrations of molecules, but for various technical reasons it had very limited value for charting the low frequencies and was not readily applicable to the kind of complex molecule of most interest in perfumery. Nevertheless, it did serve to open the vibrational theory to its first phase of systematic development.

Chapter 6

CORRELATING THE FREQUENCIES — I

The secret of all who make discoveries is to look upon nothing as impossible.

J. von Liebig

In order to correlate particular patterns of vibrational frequencies with particular odorous sensations, we need to have far infrared spectra of a number — the larger the better — of different compounds with a generally similar type of odor. The odors of different compounds are probably never *exactly* the same, but it is sometimes possible to assemble a set of compounds with the same general type of smell as evaluated by more than one judge. It is unwise to rely on a single opinion because, for reasons to be described in Chapter 24, there are enough personal idiosyncracies to make one person's judgments not altogether reliable.

The odor called "musk" is one that perfumers are very familiar with because it is distinctive and also important as a valued component of many perfume formulations. As a result, there has been an extensive program to synthesize compounds which, it was hoped, would have a musk-like odor. Many of these attempts were successful even if the products displayed by-odors or overtones, presumably as a consequence of their molecules possessing vibrational frequencies additional to those responsible for the musk pattern. In order to identify the musk frequencies we have to look for the frequencies common to a diversified assortment of chemicals with a reasonably distinctive musky-odorous character.

Figure 1 shows how the data can be analyzed. At the top left we have the far infrared spectrum of a well-known synthetic musk, musk xylol. It has absorption bands of various intensities at various places in the spectrum: 160, 180, 214, etc. wave numbers. For the moment the positions of the absorption peaks are more significant than their intensities, and so we plot their positions as dots distributed along a linear scale from 100 to 500 wave numbers (the range covered by the spectrophotometer I used in my experiments). Next, we plot on the same scale the peak positions of other compounds with a musky odor (in this case no fewer than 64 different chemicals contributed by various manufacturers of perfume chemicals). This gives the dot diagram shown, for the range 200 to 300 wave numbers at the middle of Figure 1. Evidently the dots tend to cluster at some places in the spectrum and avoid others. In order to identify the significant groupings or gaps, a card with a notch 7 wave numbers wide is moved across the dot diagram and the number of dots visible in the notch is counted for each position and plotted as shown at the bottom of the Figure. The number "7" is arbitrary, but a notch much wider would smooth out the clusters too much, and one much narrower would not smooth them out enough.

There is naturally bound to be some random rising and falling in the count so we need some criterion to decide which highs and lows are significant.

From a survey of the dot diagram for several hundred compounds presenting all sorts of unrelated odors, it was possible to estimate the number of dots to be expected in a 7 wave number interval at any given part of the spectrum by any given number of randomly selected compounds. This number is summed up by the formula

$$P_\nu = 10^{-3}\, M\, \nu^{1/2}\, d\nu$$

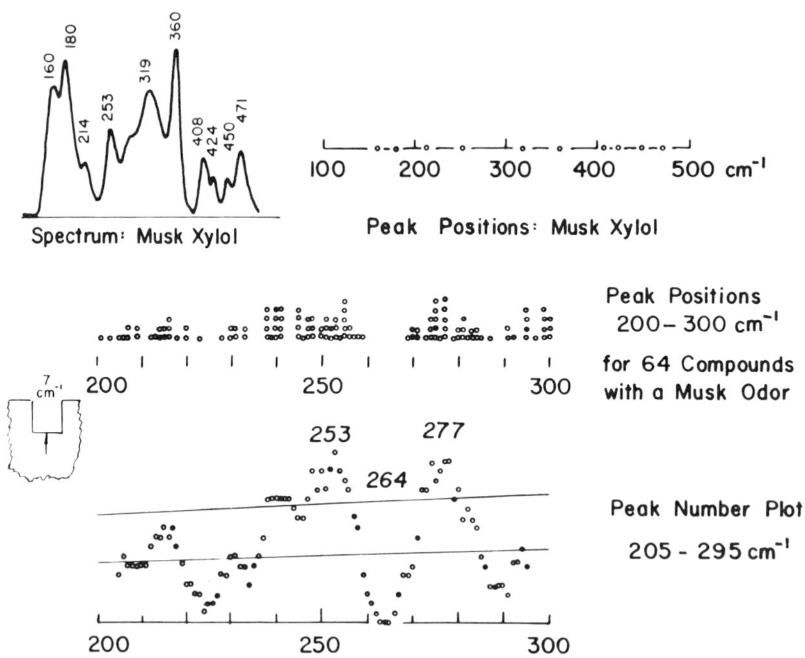

FIGURE 1. Constructing a peak number plot for substances with a musk-like odor. (From Wright, Richard H., *J. Theor. Biol.*, 64, 1977, 473. With permission.)

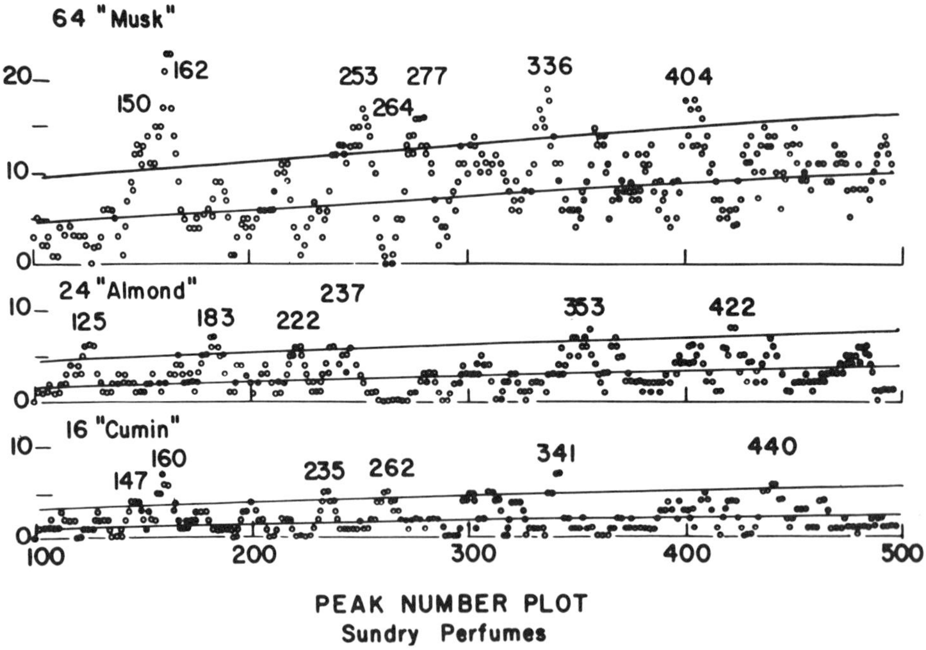

FIGURE 2. Peak number plots for substances with the musky, the bitter almond, and cumin types of odor. Each odor type correlates with a specific pattern of frequencies, which may be favorable or adverse.

Table 1
OLFACTORY PROPERTIES OF DERIVATIVES OF BENZALDEHYDE

Compound	Odorous character
Benzaldehyde	Bitter almond
o-Me Benzaldehyde	Bitter almond
o-Et Benzaldehyde	Bitter almond with a weak note of ethyl-benzene
o-iPr Benzaldehyde	Reminiscent of cumin
o-tBu Benzaldehyde	Reminiscent of t-butyl-benzene
m-Me Benzaldehyde	Bitter almond, harder note than the o-methyl
m-Et Benzaldehyde	Bitter almond, faint trace of sassafrass
m-iPr Benzaldehyde	Cumin, weak note of sassafrass
m-tBu Benzaldehyde	Cumin, strong note of carrot
p-Me Benzaldehyde	Bitter almond
p-Et Benzaldehyde	Bitter almond
p-iPr Benzaldehyde	Strong cumin
p-tBu Benzaldehyde	Cumin with a note of carrot
Nitrobenzene	Bitter almond
o-Me Nitrobenzene	Bitter almond
o-Et Nitrobenzene	Weak almond, a little like ethyl-benzene
o-iPr Nitrobenzene	Reminiscent of cumin
o-tBu Nitrobenzene	Like t-butyl-benzene
m-Me Nitrobenzene	Bitter almond
m-Et Nitrobenzene	Weak almond, trace of cumin alongside sassafrass
m-iPr Nitrobenzene	Cumin, accompanied by a weak note of sassafrass
m-tBu Nitrobenzene	Cumin, strong note of carrot
p-Me Nitrobenzene	Bitter almond (hard)
p-Et Nitrobenzene	Sassafrass, weak note of cumin
p-iPr Nitrobenzene	Cumin
p-tBu Nitrobenzene	Cumin, note of carrot

where P_ν is the number of absorption peaks to be expected in a set of M compounds selected without regard to their odor, in a narrow frequency interval, $d\nu$, in the vicinity of frequency, ν. Thus, for example, a random selection of 64 compounds would be expected to have about 7 peaks in a 7 wave number interval near 260. The line drawn through the center of the peak number plot at the bottom of Figure 1 represents this expected or average number. The line drawn above it is higher by 2 SD. This is a statistical criterion regularly used to identify deviations from a norm which are probably not the result of chance fluctuations. On this basis, vibrations near 253 and 277 appear to correlate positively with the musk odor, and one near 264 correlates negatively.

Figure 2 shows the peak number plot for the 64 musk chemicals for the entire range 500 to 100 wave numbers. Provisionally the musk pattern of frequencies can be taken to include favorable ones near 150, 162, 253, etc. wave numbers and at least one adverse frequency near 264.

Unfortunately, there are not very many readily available chemicals with the kind of easily recognized smell that musk has, but scientists with the Naarden Company in Holland made a systematic study of some compounds chemically related to benzaldehyde and found that they fell largely into two rather well-defined classes: "bitter almond" and "cumin", as shown in Table 1. The Naarden Company kindly provided me with samples of most of these, so that their far infrared spectra could be recorded, and the rest were synthesized for me by Mr. Adrian Robson. This enabled me to construct peak number plots from the spectra of 24 bitter almonds and for 16 cumins as shown in Figure 2.

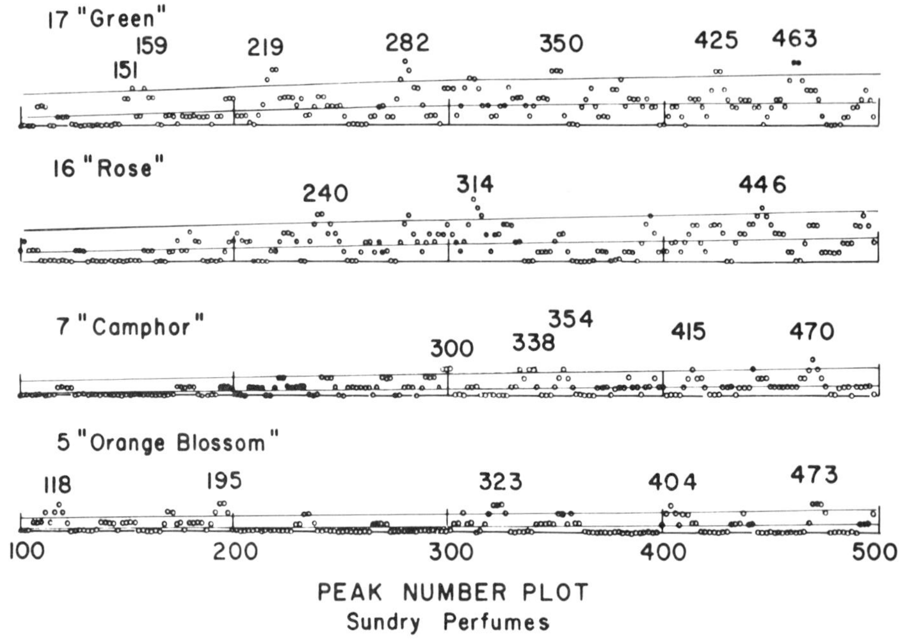

FIGURE 3. Peak number plots of varying significance for substances with the green or leafy type of odor, the rose-like odor, the camphoraceous odor and the odor of orange blossoms.

With the smaller numbers of compounds there is less certainty in deciding which are the significant frequencies, but to a fair approximation, distinctive patterns can be deduced for each type of odor. The result is just what one would have expected, based on the considerations arrived at in Chapter 2. The overall patterns are distinctive even though they may show some common or overlapping frequency-elements, such as one near 235 in the bitter almond and cumin patterns, and one near 160 in the musk and cumin patterns.

Unfortunately, the kind of careful odor evaluations using well-purified compounds such as are represented in Figure 2 are rather uncommon, so that it has not been easy to collect the data needed for the preparation of very many plots like those shown in the figure. To show the kind of difficulty one encounters, I asked several professional perfumers to evaluate the odors of toluene, and of chloro-, bromo-, and iodo-benzene in relation to the bitter-almond note of nitrobenzene. The response was unanimous that toluene does not have the almond character, but it varied for the halogenated benzenes:

Observer group	Opinion
A	Chlorobenzene has almond-like character, others unavailable
B	Almond character in all three
C—1	No almond-like character
C—2	Chloro- and bromo-, no; iodo-benzene is chemical almond
C—3	Have note similar to nitrobenzene
D	If benzaldehyde is 100, nitrobenzene is 60, the rest about 5

With so little agreement, no attempt was made to include any of these chemicals in the set of bitter almond used in plotting the peak number plot shown in Figure 2.

Subject to this sort of difficulty, it has been possible to assemble a few other sets and construct the peak plots shown in Figure 3. With the numbers of compounds and the uncertainties inseparable from this sort of work, the results are not as clear as one could wish. Nevertheless, the picture that emerges seems clear enough: each type or class of odor can be correlated more or less securely with a distinctive and specific pattern of low-frequency vibrations. What is most significant is that within each odor class there may be substances possessing the most diverse chemical constitutions and molecular shapes.

At the same time, it is necessary to mention once again the many qualifications the human odor-evaluator nearly always includes in his verbal descriptions. These are an obstacle in the way of assembling groups of substances with generally similar odors. Once such groups have been assembled, the verbal qualifications take on a new and revealing significance in ways to be considered shortly.

REFERENCES

1. **Klouwen, M. H. and Ruys, A. H.**, Chemical constitution and odor. III. Comparison of the odor of benzaldehyde and nitrobenzene derivatives with the same substituted groups, *Parfuem. Kosmet.*, 43, 289, 1962; also cited in *Parfum. Cosmet. Savons*, 6, 6, 1963.
2. **Wright, R. H. and Robson, A.**, Basis of odor specificity: homologs of benzaldehyde and nitrobenzene, *Nature (London)*, 222, 290, 1969.
3. **Wright, R. H. and Burgess, R. E.**, Musk odor and far infrared vibration frequencies, *Nature (London)*, 224, 1033, 1969.
4. **Wright, R. H.**, Molecular vibration and the "green" odor, *J. Appl. Chem. Biotechnol.*, 21, 10, 1971.
5. **Wright, R. H. and Burgess, R. E.**, Molecular coding of olfactory specificity, *Can. J. Zool.*, 53, 1247, 1975.

Chapter 7

CORRELATING THE FREQUENCIES — II

What one should then do is to supplement the other man's work with the smallest number of experiments needed to carry conviction. There is nothing I delight in more than a problem in which ninty percent of the work has already been done.

W. D. Bancroft

When human olfactory responses become too perplexing, it is a comfort to turn to the simpler and more stereotyped responses of insects. Because of their limited brain size, they have little freedom to ponder, rationalize, and choose when, whether, and how they will respond to a signal. At most, they can respond more strongly or less strongly, depending upon their physiological state, genetic makeup, and the strength and precision with which the message is presented to them. The responses are usually distinctive enough to be recorded on an all-or-nothing basis: an insect is either caught in a trap or it isn't, it flaps its wings in a certain way or it doesn't, it deposits eggs or it does not deposit them. Moreover, since insects are usually cheap to rear in large numbers, and the wild ones are available free, it is usually easier to get a statistically representative sample numbering in the hundreds than to get a dozen human evaluations.

It can, of course, be argued that the insect's smelling organs are in no way likely to be similar to those of vertebrates. There are at least two ways to answer this. The first is to say, "Shall we sit and do nothing, or shall we try it and see?". The second is to remember that we and the insects have evolved from simpler ancestors, and that far back in the process chemical senses must have been the first to appear and become genetically established.

The U. S. Department of Agriculture has carried out tests with some thousands of chemicals for the ability to attract various insects into traps. We have already made some use of the resulting data in Chapter 3. With samples of various specifically attractive compounds sent me by Dr. Morton Beroza and others of his colleagues,[1] I was able to record the far infrared spectra and construct the peak number plots shown in Figures 1, 2, and 3.

The first way to use the resulting data is to see how far it is consistent with the results shown in Tables 1, 2, and 3, Chapter 3.

Of the 15 insect pairs shown in Table 1, Chapter 3, 8 have no common frequency in their patterns, 5 appear to have 1 and 2 have 2. In most of the latter pairs there is a frequency present in one pattern that is conspicuously absent from the other. The possible significance of this is considered below. Especially interesting is the fact that the shared frequencies are the same for the triple combination shown in Table 3, Chapter 3.

These results are by no means unequivocal, but to a first approximation they do suggest the kind of correlation we are seeking. Also, taking the data in the tables and figures at their face value, there may be an implied hint that not all the contributors to a frequency-pattern are of equal importance in eliciting a response. This is something to be considered later. Meanwhile, let us see what further use can be made of the peak number plots.

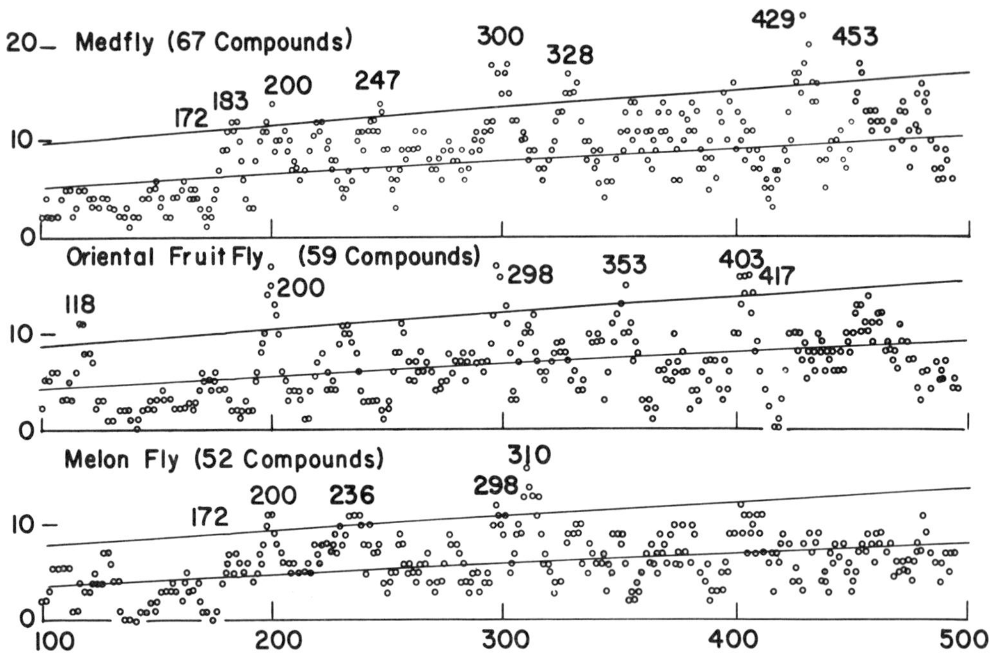

FIGURE 1. Peak number Plot. Sundry insects. (From Burgess, R. E. and Wright, R. H., *Can. J. Zool.*, 53, 9, 1247. With permission.)

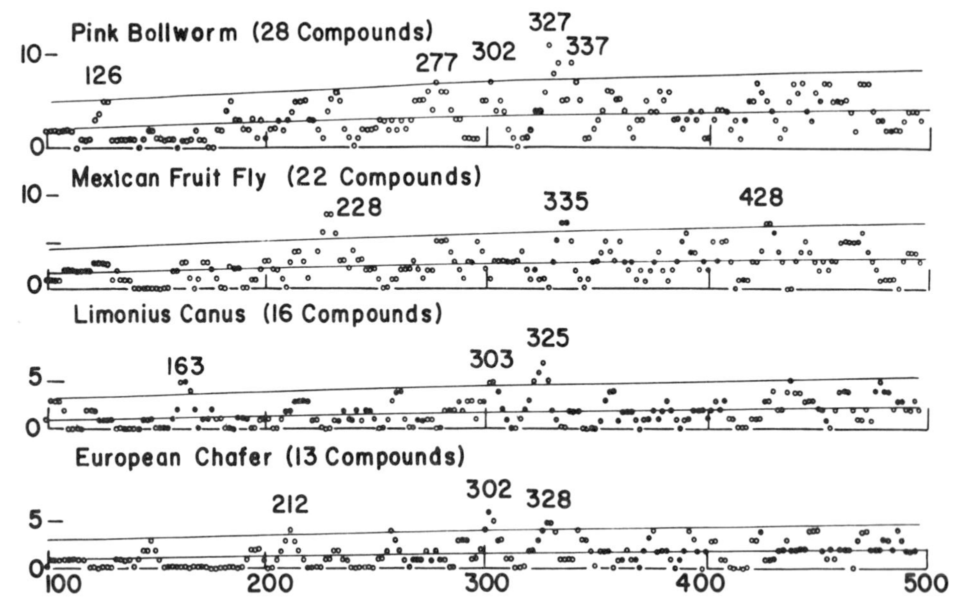

FIGURE 2. Peak number plot. Sundry insects. (From Burgess, R. E. and Wright, R. H., *Can. J. Zool.*, 53, 9, 1247. With permission.)

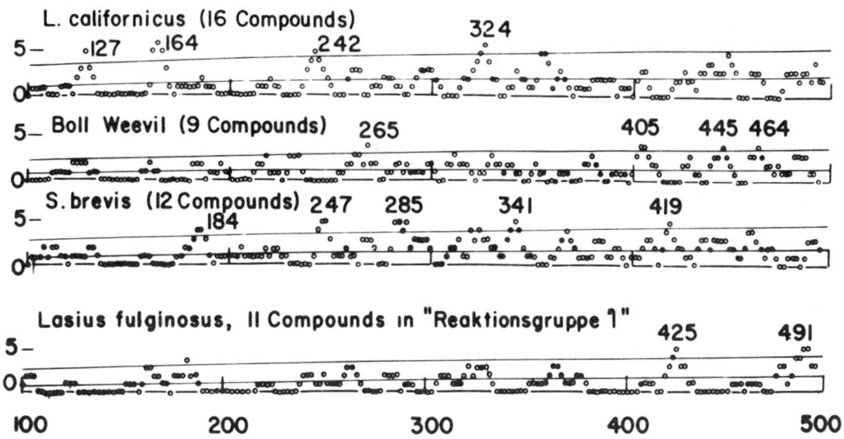

FIGURE 3. Peak number plot. Sundry insects.

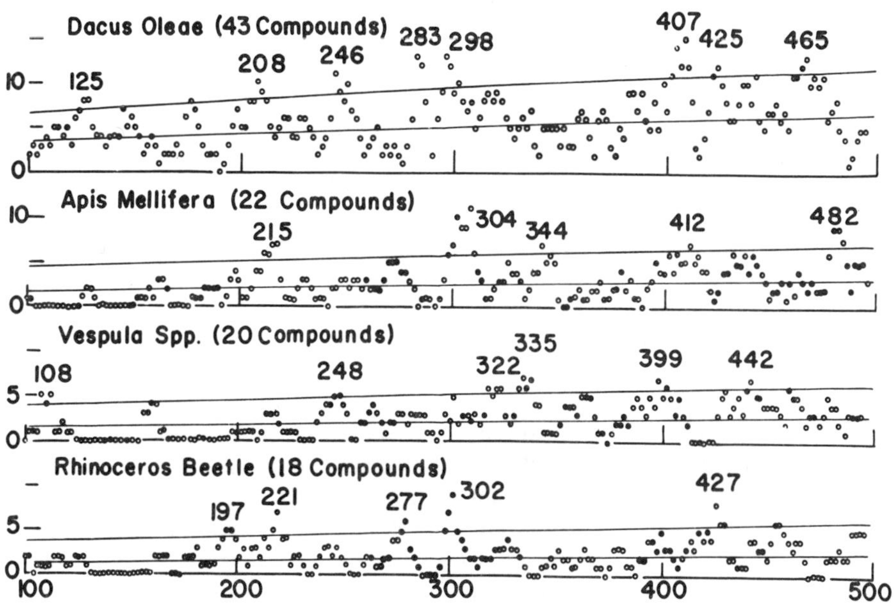

FIGURE 4. Peak number plot. Sundry insects. (From Burgess, R. E. and Wright, R. H., *Can. J. Zool.*, 53, 9, 1247. With permission.)

The most obvious and, in a sense, the most crucial test is to use them to predict attractancy in compounds before they are tested: to select candidates with a better-than-even-chance of their working. This has now been done.

Figure 4 shows peak number plots for four more species of insect.

To identify the significant frequencies for the olive fly, *Dacus oleae,* approximately 130 compounds were tested in traps in olive groves in Greece, Crete, and Corfu, and about 20 were found to be consistently if not strongly attractive. This enabled a tentative vibrational correlation to be deduced, and with its help, additional candidate attractants were selected with ever-increasing confidence. The field-tests were made by Dr. G. N. Stavrakis of the Benaki Phytopathological Institute in Athens after an initial

Table 1
MOLECULAR VIBRATION PATTERNS ENABLE CHEMICALS TO BE CHOSEN AS CANDIDATE INSECT ATTRACTANTS OR ANTIATTRACTANTS

Dacus oleae pattern	125	208	246	283	298	407	425	465
Homocysteine thiolactone		212	240	283			430	463
Diallylphthalate	120		254		294			
Rhinoceros beetle pattern		197	221	277	302	427		
N,N-diethyl-*m*-toluamide		192	217	277	300	428		
Yellow jacket wasp pattern	108	248	322	335	399	442		
Butyl benzoate				340		440		
n-Pentyl *n*-pentanoate			316	344		450		
Dacus oleae, adverse frequencies		188	265					
Menthol		189	258					
Butyl butyrate			273					
Medfly and melon fly, adverse frequency				172				
Coumarin				166				
Acetonaphthone				170				
Ethyl salicylate				174				
Butyl ether				160				

period during which the work was sponsored by the Food and Agriculture Organization of the United Nations. Among the more effective compounds found in this way were homocysteine thiolactone and diallyl phthalate, compounds with very dissimilar chemical structures but with vibrational patterns of the required sort, as shown in Table 1.

The case of the rhinoceros beetle, *Oryctes rhinoceros,* a pest of the coconut palm, is similar. Using results of field-trapping tests reported by various experimenters, I constructed a peak number plot and selected several candidates which were tested in Western Samoa by Dr. P. A. Maddison with generally positive results. One of the most surprising was that the compound, *N,N*-diethyl-*m*-toluamide, or "DEET", which is regularly used to repel mosquitoes and biting flies, is distinctly though not strongly attractive to the rhinoceros beetle. Table 1 compares its frequencies with those derived from the plot.

Yellow jacket wasps, *Vespula* species, provide a third example. During a search for compounds which might attract houseflies, H. G. Davis of the U. S. Department of Agriculture exposed some hundreds of chemicals in traps and found quite unexpectedly that a significant number of them caught yellow jacket wasps. He was good enough to send me samples of many of these attractants and I was able to record their spectra and then to predict that two previously untested compounds, butyl benzoate and *n*-pentyl pentanoate, would be effective — and they were. Both are more readily available and cheaper than the best of the previously known chemical attractants for these wasps. Their spectra included frequencies from the wasp pattern as shown in Table 1.

What makes this experiment especially significant is the remarkable specificity of the attractive effect. An examination of more than 2000 captured specimens failed to reveal any other species of insect. In another experiment, when traps were placed in the near vicinity of some beehives, some thousands of yellow jackets were trapped along with fewer than a dozen bees which had somehow blundered into the traps. The only other insect that appears to be systematically attracted along with the wasps is

Table 2
ANTIATTRACTION

Formulation	*Dacus oleae* trapped (m)	(f)
DHP alone	198	93
DHP plus menthol	1	0
DHP alone	124	47
DHP plus Butyl butyrate	7	4

Formulation	Melon flies trapped (m)	(f)	Medflies trapped (m)	(f)
Bait alone:	63	105	106	98
Bait + coumarin	12	3	74	40
Bait + acetonaphthone	37	21	182	14
Bait + ethyl salicylate	75	12	69	15
Bait + butyl ether	135	69	236	125

the earwig (Order Dermaptera). Furthermore, although the chemicals are so strongly effective in catching wasps (*Vespula* spp.) in North America, they were completely ineffective in attracting the kinds of wasps found in Australia, Israel, Switzerland, and England.

With the growing concern for environmental pollution by undiscriminating pesticides like DDT, an agent that can selectively affect one species at a time and without any toxic properties is of more than purely scientific interest — a matter to be considered more fully in Chapter 25.

It will be seen that in many of the peak number plots there are conspicuous valleys or gaps, suggesting that an insect's response may depend upon the absence of some frequencies as well as the presence of others. In selecting candidate attractants for the various insects, the adverse as well as the favorable frequencies have to be taken into account — something that is not illustrated specifically in Table 1.

The effect of adverse frequencies has been positively demonstrated so far in two experiments.

Figure 1 indicates the probability that a frequency near 172 wave numbers is adverse to the attraction of the medfly and also the melon fly. D. L. Chambers and I. Keiser with the U. S. Department of Agriculture in Hawaii added compounds with absorption maxima near 172 wave numbers to a standard lure for these insects and got the expected effect, as shown in Table 2. It is clear that an antiattractant effect exists and is greater the closer the added frequency is to 172. The fact that males and females are affected differently is interesting and worth following up, but is not strictly relevant to the immediate task which is to show that the vibrational hypothesis has the ability to generate predictions of phenomena that would not be predictable on any other hypothesis.

The second example of the antiattractant effect was recorded in experiments with the olive fly by Dr. Stavrakis. From Figure 4, there are indications that attracting compounds lack frequencies near 187 and 265 wave numbers. Butyl butyrate and menthol were added to the standard lure (an aqueous solution of diammonium hydrogen phosphate, or DHP) and once again the expected antiattractant effect was observed, as shown in Table 2.

In Chapter 3, Table 2, showed that the number of compounds found to attract both the oriental fruit fly and the boll weevil was less than expectation. This may be an example of antiattractancy due to favorable frequencies in one pattern matching adverse frequencies in the other. With the limited data available, it is not possible to identify the culprit frequencies with any certainty. For what it is worth, however, the oriental fruit fly pattern in Figure 1 has marginally significant favorable frequencies near 236 and, possibly, 454 wave numbers, which appear to match gaps in the boll weevil plot shown in Figure 3. The number of spectra used in making the peak number plot for the boll weevil is too small to permit the reverse search.

Evidently, then, the low-frequency vibrational properties of various chemicals are connected with the behavioral responses of insects in a way that is generally similar to the way humans respond to odors. (It has long been known that honeybees recognize many of the same odor similarities that we do.) We are therefore justified in accepting as a working hypothesis two propositions:

1. The basic mechanisms of smell perception are the same in vertebrates and invertebrates.
2. The low-frequency vibrational attributes of molecules correlate with the odorous sensations the molecules elicit.

These propositions relate to the intrinsic type of smell or insect response. The strength of the smell or the number of insects which respond will undoubtedly depend upon other, unrelated molecular attributes of which volatility is the most obvious. There is no way the molecules of a completely nonvolatile compound can reach the odor receptors of an air-breathing creature. Even if a compound is volatile, the number of molecules which actually reach the receptors will depend also upon the physical conditions of the experiment, including the geometry of the smelling organ. Finally, the nature of the interaction between stimulus and receptor may make it easier for some molecules than others to elicit a response.

Subject to these qualifications, the propositions as stated appear to be reasonably secure on the basis of the evidence presented. Nevertheless, there is still much to be found out. For example, I ran the spectrum of 9-oxo-2-decenoic acid which is the sex-attractant scent used by the queen honeybee, *Apis mellifera,* to attract the drones. However, when scientists at the Rothamstead Experimental Station in England made tests with several compounds with the appropriate spectral frequencies, the results were negative in every case. This at first disappointing result is still as much a fact as the previously described predictive successes, and like them, it is telling us something. What it may be saying will be discussed later, in Chapter 21.

REFERENCES

1. **Beroza, M. and Green, N.,** Materials tested as insect attractants, U. S. Department of Agriculture Handbook No. 239, Washington, D. C., 1963.
2. **Blum, M. S., Warter, S. L., and Traynham, J. G.,** Chemical releasers of social behavior. VI. The relation of structure to activity in ketones as releasers of alarm for *Iridomyrmex pruinosus* (Roger), *J. Insect Physiol.,* 12, 419, 1966.
3. **Boch, R. and Shearer, D. A.,** Chemical releasers of alarm behavior in the honey bee, *Apis mellifera, J. Insect Physiol.,* 17, 2277, 1971.
4. **Burgess, R. E. and Wright, R. H.,** Evidence for a common mechanism of smell stimulation in man and insects, *Chem. Senses Flavor,* 1, 89, 1974.

5. **Davis, H. G., McGovern, T. P., Eddy, G. W., Nelson, T. E., Bertun, K. M. R., Beroza, M., and Ingangi, J. C.**, New chemical attractants for yellow jackets (*Vespula* spp.), *J. Econ. Entomol.*, 61, 459, 1968.
6. **Lehman, R. S.**, Experiments to determine the attractiveness of various aromatic compounds to adults of the wireworm *Limonius* (Pheletes) *canus* Lec. and *Limonius* (Pheletes) *californicus* Mann, *J. Econ. Entomol.*, 25, 949, 1932.
7. **Stavrakis, G. N. and Wright, R. H.**, Molecular vibration and insect attraction: *Dacus oleae*, *Can. Ent.*, 106, 333, 1974.
8. **Wright, R. H. and Brand, J. M.**, Correlation of ant alarm pheromone activity with molecular vibration, *Nature (London)*, 239, 225, 1972.
9. **Wright, R. H.**, Methods and Compositions for Selectivety Attracting Yellow-Jacket Wasps, Canadian Patent, 900,257, 1972. U. S. Patent, 3,798,318, 1974.
10. **Wright, R. H.**, Predicting olfactory quality from far infrared spectra, *Ann. N.Y. Acad. Sci.*, 237, 129, 1974.
11. **Wright, R. H., Chambers, D. L., and Keiser, I.**, Insect attractants, anti-attractants and repellents, *Can. Entomol.*, 103, 627, 1971.
12. **Wright, R. H., Chapman, J. A., and Dyer, E. D. A.**, Molecular vibration and insect attraction: *Dendroctonus rufipennis*, *Bi. Mon. Res. Notes For. Serv., Environ. Can.*, 30(2), 10, 1974.
13. **Wright, R. H. and Burgess, R. E.**, Molecular coding of olfactory specificity, *Can. J. Zool.*, 53, 1247, 1975.
14. **Dumpert, K.**, Receptors for alarm substances on the antenna of *Lasius fulginosus* (Latr.), *Z. Vgl. Physiol.*, 76, 403, 1972.

Chapter 8

REDUNDANCY

English text, and most other information sources are *redundant* in that the messages they produce give many clues to the recipient. A few errors caused by replacing one letter by another don't destroy the message because we can infer it from other letters which are transmitted correctly. Indeed, it is only because of this redundancy that anyone can read my handwriting.

J. R. Pierce

Inspection of the peak number plots makes it plain that few chemical molecules are likely to present the whole frequency-pattern for a given sensation. If various subsets of the pattern can in some degree act for the whole, it means that there is redundancy built into the system. Just how small a part of the pattern can serve as a surrogate for the whole can be estimated as follows. Figure 2, Chapter 6 shows six favorable frequencies for the musk pattern. With spectra of 67 compounds reported to have some degree of muskiness in their odors, it is possible to count the number of peaks within eight units of the indicated frequency in each of the compounds. The results are shown in Table 1. From this it appears that three musk frequencies are enough to confer a distinctly musk-like odor. With only two, the odor is likely to be weak and blended with a nonmusk by-odor, and with only one the musk character is weak at best.

It is to be noted also that there is some uncertainty in the precise position of a far infrared peak and some solvent effect on the precise position of the absorption maximum. Table 1 uses the same frequencies as those used in constructing the peak plots.

A similar comparison can be made to establish the approximate amount of redundancy in the bitter-almond pattern. Figure 2, Chapter 6, shows six favorable frequencies, and Table 2 would indicate that the amount of redundancy is about the same as for the musk odor.

Similar results are revealed by the corresponding scrutiny of insect responses. Once again, it is unlikely that very many compounds will present the whole of the attractancy pattern derived from the peak number plot for a given insect species. The medfly pattern in Figure 1, Chapter 7, with seven potentially favorable frequencies can be expected to include a substantial amount of redundancy, and this expectation is fully confirmed by Table 3. It is, however, necessary to look rather closely at the meaning given the word "attractancy" by the U. S. Department of Agriculture scientists who assembled the bioassay data.

The materials were classified according to an "attractancy index" which was obtained by dividing the number of insects caught with the candidate lure by the number caught with plain water. For medflies, the three attractancy classes were

	Attractancy Index for	
Class	**Males**	**Females**
1	Less than 11	Less than 6
2	From 11 to 50	From 6 to 50
3	Greater than 50	Greater than 50

Table 1
REDUNDANCY IN THE MUSK PATTERN

Number of musk frequencies	Number of compounds	Perfumers' prescriptions
5	2	(1) Weak but pure at 150°C (2) Galaxolide®![a]
4	17	Mostly weak to moderate with mention of other notes; the group includes Exaltolide, Moskene, and Versalide
3	23	A few weak, mostly medium, some strong; the group includes Musk 89, Exaltone, Musk Ketone, Musk Xylol, and Musk Tibetene
2	16	Mostly weak, some moderate and two strong with additional character, including Musk Ambrette and Celestolide (a very strong musk with the fatty character of angelica seed)
1	6	Five weak, one intermediate
0	3	One weak, the other two are Astratone and Civettone which are macrocyclics with rather poor spectra.

[a] A trade name is evidence of marketability.

Table 2
REDUNDANCY IN THE BITTER-ALMOND PATTERN

Number of favorable frequencies	Number of compounds with almond odor
5	2
4	0
3	6
2	7
1	8
0	1

Table 3
REDUNDANCY IN THE MEDFLY PATTERN

Numbers of favorable frequencies	Number of compounds	Attractancy class
7	1	Class 2
6	2	Both Class 3, one is Trimedlure[a]
5	4	One Class 3, three Class 2
4	12	Four Class 3, eight Class 2
3	17	Five Class 3, twelve Class 2
2	15	Two Class 3, thirteen Class 2
1	11	Four Class 3, seven Class 2
0	3	All Class 3

[a] Trimedlure is one of the most effective Medfly attactants for practical use.

Table 4
VIBRATIONAL PATTERNS OF 100 NONMUSKS

Number of musk-favorable frequencies	Number of compounds	Number with musk-adverse frequencies	Number of musk-mimics
4	2	0	2
3	8	1	7
2	28	6	22
1	36	13	23
0	26	6	0

In compiling the peak number plot, only chemicals in Classes 2 and 3 were considered. It is evident that compounds in Class 1 might have some attractant effect but not enough to be practically interesting. Furthermore, it is clear that a given lure may affect males and females differently, but the attractancy ratings published by the U. S. Department of Agriculture do not discriminate on the basis of sex.

At this point, the alert critic will point out that if as few as two elements of the musk pattern arc sufficient for the musky odor to be registered, then it should be possible for many compounds with no musky odor to meet this reduced vibrational specification. The correctness of this estimate is borne out by a scrutiny of the spectra of a random selection of 100 nonmusk chemicals. The results are shown in Table 4.

At first sight, this appears to be an unanswerable objection to the vibrational theory, and it is indeed unanswerable if it be held that the possession of one or two musk-favorable and no musk-adverse frequencies is a *necessary and sufficient* condition for the musk odor to be perceived. As a quantized transfer of energy between structures of molecular size, more is required than a simple matching of the frequencies. Not the least important of these additional factors will be the relative orientation of the interacting dipoles. This will depend upon which of its several faces or profiles the stimulus presents to the receptor at the moment of interaction, and also upon the way the various parts of the molecule move to-and-fro when it becomes vibrationally excited as a result of the interaction.

With vibrational frequency determining the *possibility* of a stimulus-receptor interaction and molecular shape or profile determining the *probability,* what have hitherto been held to be mutually exclusive rival theories of olfaction are now seen to be mutually complementary. Each fills out the gaps in the other, and both must eventually be invoked before the details of the stimulus-receptor interaction can be worked out. The point will be raised again in later chapters.

Meanwhile, the recognition of the existence of redundancy in the channel throws an interesting light on some matters touched on in Chapter 3. The statistics of insect attractancy of various chemicals compiled by the U. S. Department of Agriculture showed that of 2577 assorted chemicals tested as medfly attractants, 652 or 25% were effective. With Figure 1, Chapter 7, showing seven potentially favorable frequencies and with as few as two being potentially capable of evoking a response, it is not surprising that almost any randomly selected compound has a 1:4 chance of being appreciably attractive. When a comparable number of compounds, 2618, were assayed of the ability to lure the Mexican fruit fly, only 205 or 8% were effective. Figure 2, Chapter 7, shows only three potentially favorable frequencies for this insect, making it less likely that a randomly selected compound will present one or two.

Given the fact of redundancy and the different factors that govern the possibility as distinct from the probability of a given odor being perceived, many otherwise obscure matters begin to make sense, and the way is opened to understanding some of the hows and whys of odorous complexity.

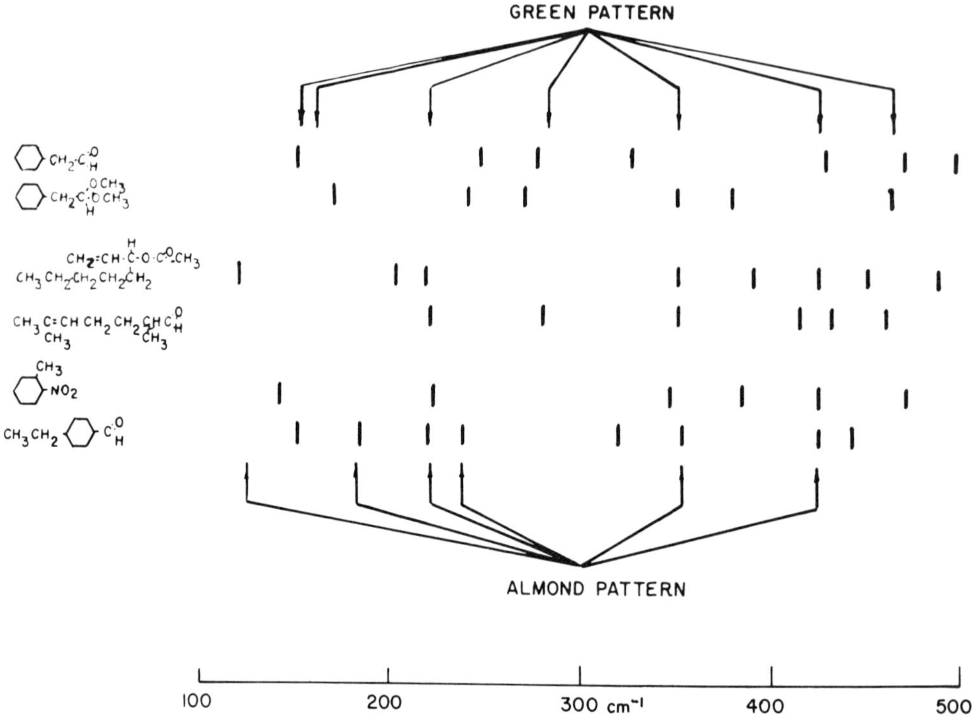

FIGURE 1. The first two compounds have a green type of odor which is consistent with the possession of several frequencies in the green pattern. Similarly, the last two have frequencies present in the almond pattern sufficient to account for their bitter-almond odor. Simply on the basis of their infrared frequencies, the middle pair could be either green or almond. In fact, they are predominantly green, presumably because stereochemical factors help to characterize the osmically active molecular vibrations.

Thus, for example, the chemical, Rose Oxide, or 2-(2-methyl-1-propenyl)-4-methyl-tetrahydro pyrane, was evaluated for rose, green, and other notes by six expert observers with the following results.

Observer	Rose?	Green?	Other?
A	Weak	No	Geranium
B	+		Geranium
C	None	None	Naphthalenic
D	Strong	Strong	
E	Part	Part	
F	None	Medium	Strong peppery, weak flowery

This compound has far infrared peaks at the following frequencies: 100, 172, 227, 284, 314, 372, 412, 425, and 470 cm^{-1}. One of these, 314, coincides with an element in the rose pattern (Figure 3, Chapter 6), and two at 284 and 425 are close to peaks in the green pattern. Its peaks at 172, 372, and 470 are fairly close to peaks in the spectrum of naphthalene at 180, 360, and 475 cm^{-1}.

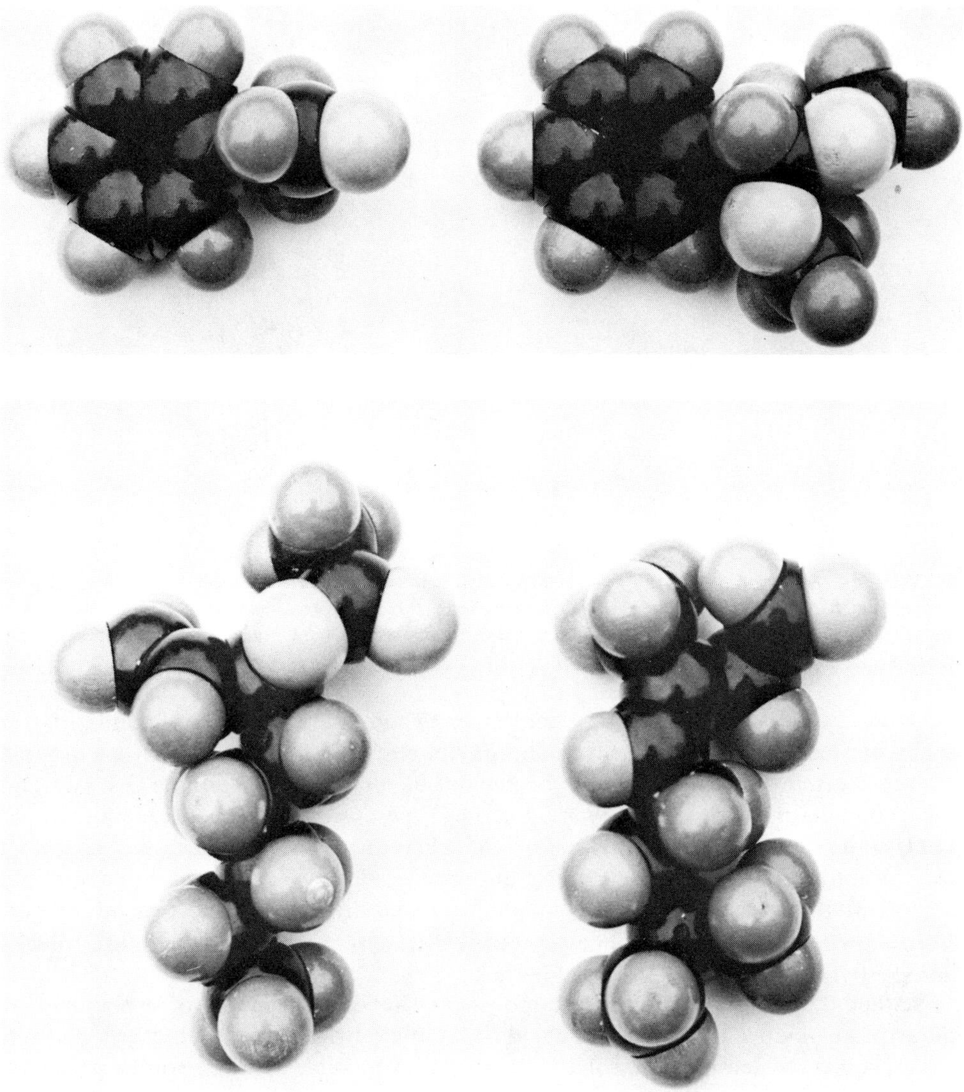

FIGURE 2. Upper: models of phenyl acetaldehyde and its dimethyl acetal; lower: models of amyl vinyl carbinyl acetate and of 2,6-dimethyl-5-heptenal. All four have a green type of odor despite their considerably different shapes. It is probable that a sufficiently detailed knowledge of their Normal Mode vibrations coupled with an understanding of their orientation with respect to the receptor would bring together the stereochemical and vibrational aspects of the problem.

With human evaluations presented in words, this sort of complexity is normal. The following compounds have a predominantly green type of odor as shown by several expert evaluations.

1. Phenyl acetaldehyde
2. Phenyl acetaldehyde dimethyl acetal
3. Amyl, vinyl carbinyl acetate
4. 2,6-Dimethyl-5-heptenal

Figure 1 shows that numbers 3 and 4 have vibrational frequencies that fit as well into the almond pattern as into the green. It is evident that on purely vibrational grounds they would be as likely to possess the almond as the green type of odor.

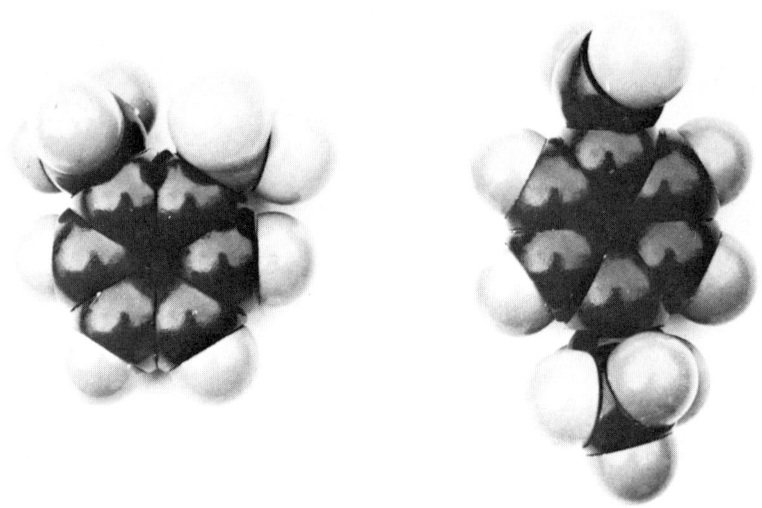

FIGURE 3. Models of *o*-nitro toluene and of *p*-ethyl benzaldehyde, both of which have a bitter-almond type of odor. Elements of their vibrational patterns appear to fit the green pattern as well as the almond pattern. It is probable that stereochemical factors determine which of the vibrational modes predominate in sending a signal or message into the organism.

Figure 2 shows models of these molecules. There are some general similarities in shape between phenyl acetaldehyde and its dimethyl acetal, and between amyl, vinyl carbinyl acetate and 2,6-dimethyl-5-heptenal, but very little between the first pair and the second despite their similarity in odor. The general similarities between pairs would imply some similarity in the way they would orient themselves with respect to an olfactory receptor and hence in the relative importance of their various vibrational modes as potential neural excitants. They might very well differ profoundly in this respect from the vibrationally similar but olfactorily different bitter almonds, *o*-nitro toluene and *p*-ethyl benzaldehyde shown in Figure 3.

Neither the spectral data nor the molecular shapes are capable of explaining the odors of amyl vinyl carbinyl acetate and of 2,6-dimethyl-5-heptenal, but they do present a problem which must be solved eventually. A possible solution will be presented in the next chapter.

REFERENCES

1. **Amoore, J. E.,** *Molecular Basis of Odor,* Charles C Thomas, Springfield, Ill., 1970.
2. **Beets, M. G. J.,** *Structure-Activity Relationships in Human Chemoreception,* Applied Science Publishers, Barking, Essex, England, 1978.
3. **Muller, A.,** Dipolar theory of the sense of odor, *Perfum. Essent. Oil Rec.,* 27, 202, 1936.
4. **Wright, R. H.,** Odor and molecular vibration. Redundancy in the olfactory code, in *Odor Quality and Chemical Structure,* Moskowitz, H. R. and Warren, C. B., Eds., A. C. S. Symposium Series, 148, American Chemical Society, Washington, D.C., 1981, 123.

Chapter 9

CORRELATING THE FREQUENCIES — III

At such times there is especial need for a statement of the guiding ideas which shall give instruction to the student and advice to the investigator.

W. Nernst

The far infrared spectrum of a compound as recorded with the sample dissolved in a solvent like benzene shows approximately which frequencies the compound may present to an organism's receptors. The indications are no more than approximate for the following reason.

Depending upon the solvent used, the frequencies derived from the positions of the absorption maxima of the solution are usually shifted from the positions seen with the compound in the vapor state. Thus, with naphthalene we find:

Frequencies in the vapor:	166.8	359.6	473.8 cm^{-1}	(wave numbers)
Frequencies in benzene:	181.8	360.7	477.0 cm^{-1}	

Evidently the close proximity of other molecules distorts the structure and perturbs the frequencies to some extent. It would doubtless be desirable to record the spectra of odorous substances in the vapor state, but this is something that is only now beginning to be technically feasible.

Even so, the interaction between stimulus and receptor would be likely to involve some degree of perturbation depending upon the closeness of approach. The interaction can be pictured as taking place in three possible ways. First there is the fly-by model in which the stimulus waves to the receptor as it goes by and without stopping or ceasing to rotate. Second, there is the touch-and-go model in which the stimulus molecule is physically adsorbed on the receptor for a brief moment during which its rotation is halted but its vibrations are not radically altered. Third, there is the hit-and-stick model in which the stimulus falls into a relatively deep potential well and is held for a long period by something approaching chemisorption which would so alter the vibrational frequencies as to render impossible the kind of correlations we actually observe.

Thus the touch-and-go model is inherently the most probable, and some uncertainty in the frequencies which the stimulus actually presents has to be accepted. This point will be raised again in Chapter 15.

Given that the far infrared spectrum of a compound shows approximately the frequencies that compound may present to the organism's receptors, it still cannot tell us to what frequencies the various types of receptor are responsive, nor which combination of receptors must be stimulated for a particular sensation to be generated. The peak number plots are the means by which we get our first prospect of answering these questions. The positions of the maxima (or minima) in these plots can, however, give no more than an approximate indication of the receptor frequencies. This is because the specific stimulus frequencies depend upon the particular stimuli chosen, and the response depends upon certain of these frequencies being close to the receptor frequency but not necessarily precisely coincident with it. The stimulus frequencies will cluster around that of a receptor, but they will not necessarily cluster symmetrically, and so the center of a cluster will not usually coincide with the precise frequency of a

Table 1
SPACING OF THE MEDFLY FREQUENCIES

Frequency (from peak number plot)	Difference
172	11
183	17
200	47 = 11.8 × 4
247	53 = 13.2 × 4
300	28 = 14 × 2
328	101 = 12.6 × 8
429	24 = 12 × 2
453	

Table 2
FREQUENCY SEPARATIONS IN THE ORIENTAL FRUIT FLY

Frequency (from peak number plot)	Separations				
118	82				
200	180	98			
298	235	153	55		
353	285	203	105	50	
403	299	217	119	64	14
417					

receptor. Nevertheless, to a first approximation they offer an invaluable opening through which to penetrate the secret of the receptor mechanism.

While examining the peak number plots, Professor R. E. Burgess of the University of British Columbia remarked that the indicated receptor values seemed to be spaced out at intervals which were approximately whole multiples of a basic interval of about 12 cm^{-1}. Thus, for example, the medfly frequencies shown in Figure 1, Chapter 7, are spaced as shown in Table 1.

In order to verify the reality of the effect, the following method was devised.

In Table 2, the wave number separation between each pair of frequencies for the oriental fruit fly is shown. Similar tabulations were made for the seven peak number plots shown in Chapter 6 and the 15 plots shown in Chapter 7. A dot diagram was then prepared showing how the differences are distributed along a linear range of slightly over 200 wave numbers. There are, to be sure, differences greater than this, but the number is too small to be useful.

A peak difference plot was prepared in the same way as were the peak number plots, except that a card with a notch only 3 wave numbers wide was moved across the dot diagram. Figure 1 shows the resulting plot. The striking periodicity it demonstrates served to verify the previous rather qualitative impression: the biologically significant receptor frequencies, F, form a linear sequence of the form,

$$F = aN + b$$

against a whole number, N, with certain frequencies absent from any given pattern.

Furthermore, from Figure 1 the constant, a, can be determined with considerable

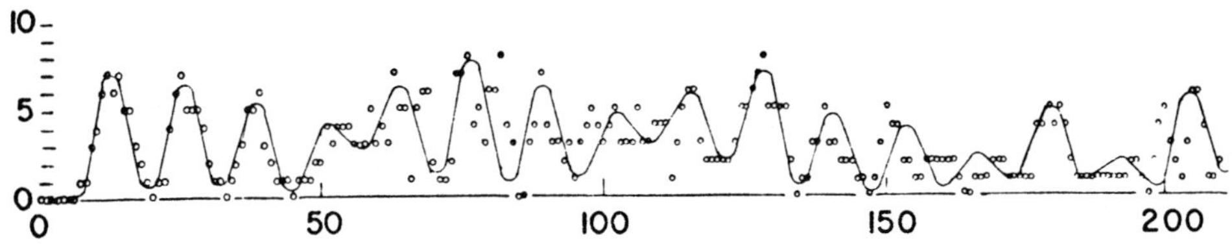

FIGURE 1. The plot of differences between frequencies associated with each of the odor categories exhibits a periodicity and shows that the osmically significant frequencies tend to be separated by whole multiples of 12.8 wave numbers. (From Burgess, R. E., and Wright, R. H., *Can. J. Zool.*, 53, 9, 1247. With permission.)

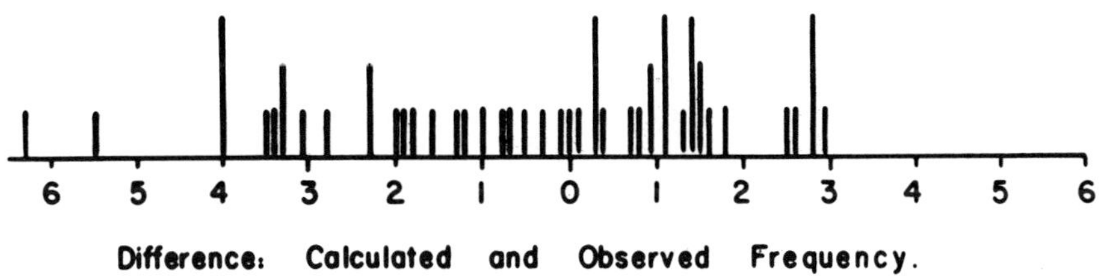

FIGURE 2. Receptor frequencies deduced from the peak number plots are grouped in a statistically significant manner around the values calculated from the formula, F = 6.4N′.

precision, and has a value of 12.8 wave numbers. For the second constant, b, in the formula the best value based on all the presently available data is 6.4 wave numbers. This is just half the coefficient, 12.8, so that the formula reduces to,

$$F = 6.4N'$$

where N′ is now an *odd* integer between about 15 and about 81.

To estimate the statistical significance of this result, it can be calculated that if the frequencies revealed by the peak number plots were simply random numbers, their root-mean-square (rms) deviation from values computed from the formula would be 3.7 cm^{-1}. The actual rms deviation is 2.8. This result is also shown graphically in Figure 2 where the values of the receptor frequencies from all the plots are grouped around the nearest calculated values instead of being spread evenly as they would be if they were simply random numbers.

The existence of a common value for the coefficient, a, for the different species, human and insect, is established independently of any assumption about b because a frequency difference has the form $F_2 - F_1 = a(N_2 - N_1)$. This, together with the fact that b also has a common value is a cogent indication that the receptor mechanism is essentially the same among the species.

Bearing in mind the facts of evolution, this is not an unreasonable conclusion. The chemical constitution of the visual receptors is basically similar across the zoological spectrum, and the same is true of respiratory pigments, enzymatic and neural mechanisms, and, most of all, the molecular encoding of genetic information. Once having found a good solution to a problem, nature tends to develop and elaborate it without changing its essential operating principle except for the best of reasons. Such a reason is illustrated by the contrast between the vertebrate and the insect eye. Structurally, they differ enormously because the wave-like properties of light make it impossible to construct a lens capable of forming an image and at the same time small enough to be mounted in the head of a mosquito. Hence the structures and mechanisms whereby rays of light are directed onto the optic nerves must vary with the size of the organism, but the molecular-scale chemistry involved in translating light quanta into nerve signals can be pretty much the same in a mosquito as in an elephant.

As a molecule-to-molecule interaction, presumably a quantum interaction, olfactory stimulation has not been subject to the kind of scaling difficulty that arises in constructing organs of vision and hearing, and so has been able to continue essentially unchanged from time immemorial so far as the primary interaction mechanism is concerned. Where evolution has operated to produce truly amazing elaborations is in the information-processing mechanisms in the Central Nervous System (CNS).

For the moment, if it does nothing else, the evidence presented here for a basic similarity in the olfactory receptor mechanisms of vertebrates and invertebrates widens enormously the number of experimental approaches open to the researcher. Some of these have practical applications to be discussed farther on. What matters here is the curious fact that the receptor frequency-sensitivities are evenly spaced.

It is a fact, and a fact that is trying to tell us something.

REFERENCES

1. **Burgess, R. E. and Wright, R. H.,** Evidence for a common mechanism of smell stimulation in man and insect, *Chem. Senses Flavor,* 1, 89, 1974.
2. **Burgess, R. E. and Wright, R. H.,** Molecular coding of olfactory specificity, *Can. J. Zool.,* 53, 1247, 1975.

Chapter 10

OLFACTORY THRESHOLDS

When a chemist is fortunate enough to perceive one ray of light penetrating the dark region of organic nature, which may mark the entrance to the right path of future knowledge, he has reason to feel encouraged, although fully conscious of the vastness of the field which lies before him.

J. von Liebig

There is an astonishing disparity in our sensitivity to different odorous chemicals. Human thresholds range from as low as 5.76×10^{-14} g molecules/ℓ of air for toluene musk up to 6.31×10^{-5} mol/ℓ for ethane. A billion-to-one difference in stimulatory efficiency is another fact that is trying to tell us something.

The words "stimulus" and "receptor" seem to imply that the role of the stimulus is to give something (such as an energy quantum) *to the receptor*. In fact, the transfer could equally well be in the opposite direction, and if, thereafter, the stimulus molecule must transfer the carried-away energy to other molecules in the environment before it can go back and take another quantum from another site, the turnaround time will largely determine its stimulatory efficiency. If the ability of a stimulus molecule to interact with a receptor cell depends upon its ability to carry away a quantum of energy from the receptor, the molecule must not only have a vibrational frequency matching that of the site but it must also be in a de-excited state when it arrives at the site.

Lambert and Salter at Oxford used ultrasonic dispersion and absorption measurements to find the average number of intermolecular collisions required for a vibrationally excited molecule to lose this kind of extra energy. To understand what they did, remember that in an intermolecular collision all the forms of energy are pooled and redistributed in a random fashion among the translational, vibrational, and rotational degrees of freedom when the molecules separate. Lambert and Salter used a simple but ingenious method to bring about an unusually large number of high-energy collisions in a gas, without simply applying an external source of heat to it. When sound passes through a gas it is transmitted in the form of longitudinal or push waves — alternating condensations and rarefactions such as commonly arise from the in-and-out movements of the diaphragm of a telephone receiver or radio loudspeaker.

In the process of being pushed together in the condensation part of the sound wave, there is a local bashing together of the molecules so that they collide extra often and with extra violence. Normally, the effect is too small to make much difference, but if very high-frequency sound (ultrasound) is used, there are so many condensations and so much bashing together that an appreciable number of molecules are given an excess of vibrational energy over what they would have at the prevailing temperature.

These molecules will make many collisions with more ordinary molecules before the next ultrasound wave comes along, and eventually their extra vibrational energy will be lost to other molecules, mainly by giving them extra velocity of translation. Part of the energy put in as sound will then appear as heat because a gas warms up when the average kinetic energy of its molecules is increased. This will happen, however, only if the sound is sent through a gas which contains molecules which can, in fact, vibrate. With simple gases like oxygen or nitrogen, the stiffness of the bond in the O=O or N=N molecules is so great that there is very little chance of their becoming vibrationally excited, though they can rotate and usually do. Therefore they show rather small tendencies to absorb ultrasound by the mechanism just described.

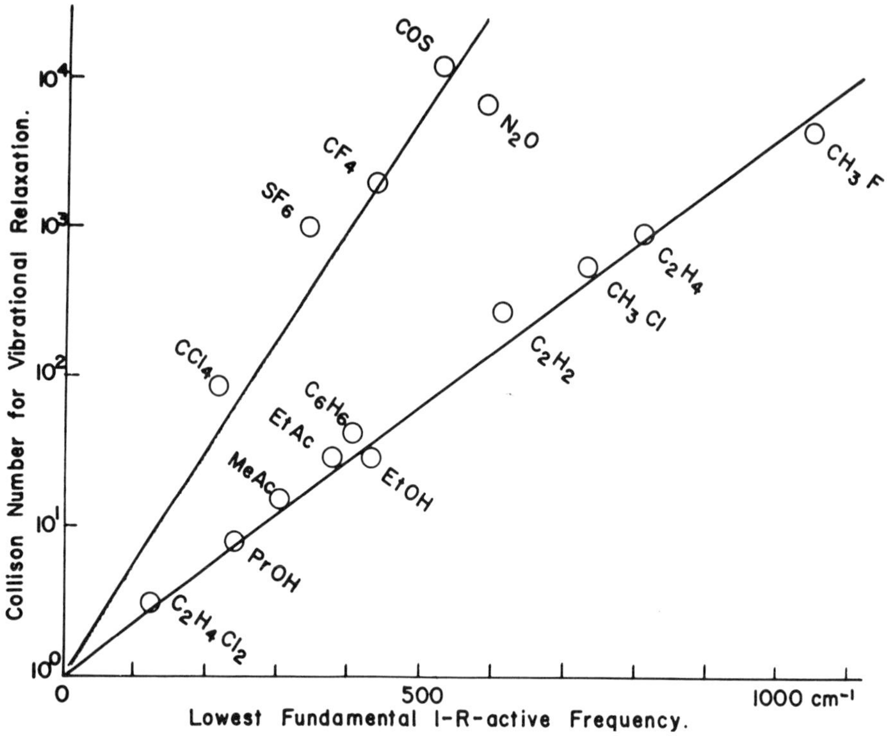

FIGURE 1. The relation between the "collision life" of vibrationally-excited molecules and the lowest fundamental frequency, ν min.

If more complex, polyatomic molecules which can vibrate are mixed with the air, then they may be vibrationally excited and subsequently pass on the extra energy to other molecules by Vibration-to-Translation (V-T) energy transfer.

The Lambert and Salter measurements using ultrasound enabled them to measure *the average number of intermolecular collisions for a V-T transfer.* This number, which they called the "collision number for vibrational relaxation" turned out to depend mainly on two things, of which the most important was the *lowest vibratonal frequency* observable in the far infrared absorption spectrum of the molecule, or ν_{min}. A second factor of some importance was the presence or absence of hydrogen atoms in the molecule, and a third, relatively minor complication arose from Vibration-to-Rotation (V-R) transfers.

Their results are summarized as two lines on a graph, the Lambert-Salter Plot shown in Figure 1. For molecules with no hydrogen atoms, the logarithm of the collision number for V-T relaxation, Log Z_{10}, gives a straight line when plotted against ν_{min}, and an equally straight line with a slightly different slope is given by molecules containing two or more hydrogen atoms.

There was no ready theoretical reason why the vibrational relaxation should depend upon the lowest vibrational frequency, but their experiments using more than 50 assorted polyatomic molecules establishes it as a fact of empirical observation.

An equally empirical fact of observation was made by Professor R. E. Burgess[2] who plotted the logarithm of the olfactory threshold concentrations of various compounds against their ν_{min}s, and found the same straight-line relation, as shown in Figure 2.

Putting the two things together, we arrive at a very simple explanation for the enormous disparity in the stimulatory effectiveness of various odorous compounds: the

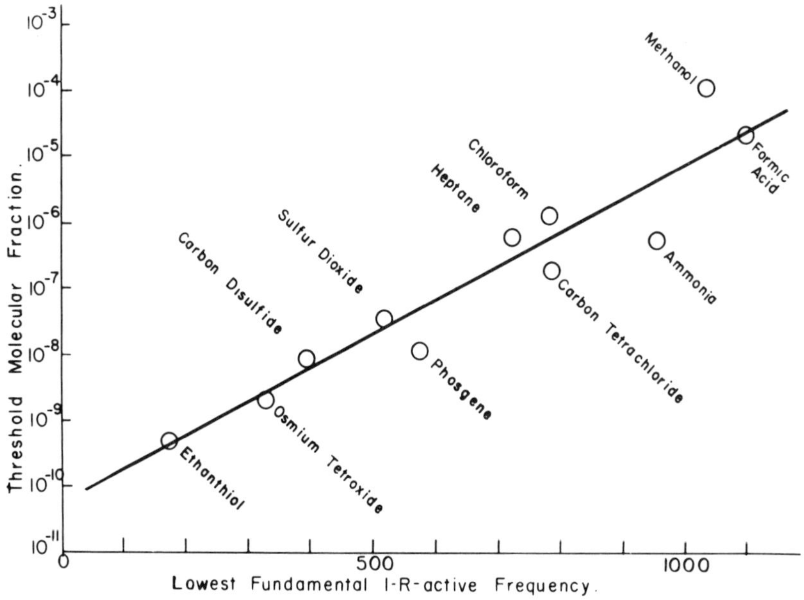

FIGURE 2. The straight-line plot of lowest IR-active frequency against olfactory threshold concentration presents an interesting parallel to the Lambert and Salter plot of lowest IR-active frequency against the probability of a collision resulting in an interchange of vibrational and translational energy.

stimulatory efficiency correlates directly with the minimum time that must elapse after a molecule has interacted with one receptor site and before it can interact with a second one.

The situation now is this: the process of interaction between stimulus and receptor seems to be one in which the special vibrational attributes of the stimuli enable them to take vibrational energy from the metabolically excited receptors and pass it into other molecules as translational energy. The ultimate receivers of the energy are unconnected with the receptors, and the receptors are destabilized as a result of their seemingly insignificant energy loss. Notice, too, that the molecules of the stimulus are in no way used up or taken out of circulation (except momentarily) in the process. Anything in the nature of a hit-and-stick interaction mechanism must surely be inconsistent with the ability of fewer than 50 stimulus molecules to engender an odorous sensation.

REFERENCES

1. **Lambert, J. D.**, Vibration-translation and vibration-rotation energy transfer in polyatomic molecules, *J.C.S. Faraday Trans. II*, 68, 364, 1972.
2. **Wright, R. H. and Burgess, R. E.**, Specific physicochemical mechanisms of olfactory stimulation, in *Taste and Smell in Vertebrates*, Ciba Foundation Symp., J. and A. Churchill, London, 1970, 325.
3. **Wright, R. H.**, Odor and molecular vibration: a possible membrane interaction mechanism, *Chem. Senses Flavor*, 2, 203, 1976.
4. **Laffort, P.**, Attempted standardization of olfactory thresholds of humans for 192 pure compounds, *Arch. Sci. Physiol.*, 17, 75, 1963.

Chapter 11

INTERACTION TIME

Bodies which are either absolutely hard, or so soft as to be void of elasticity, will not rebound from one another.

Isaac Newton

The 12.8 cm^{-1} separating the receptor frequencies means that a stimulus molecule with a normal mode frequency of, say, 200 cm^{-1} cannot interact with a receptor tuned to 187.2 or 212.8 cm^{-1}. It is reasonable to suppose that it cannot interact because it and the receptor cannot march in step long enough for the quantized interaction to take place. We can estimate roughly how long they keep step in the following way.

If the stimulus is an oscillator with a frequency, a, and the receptor one with a frequency, b, cycles per second, they will both pass through their equilibrium positions (be in phase) at a certain instant. They will once again be in phase when the faster oscillator has gained one complete cycle on the slower one. If this interval is T, the faster oscillator will make aT oscillations while the slower one makes bT + 1, so that

$$aT = bT + 1$$

and

$$T = \frac{1}{a - b}$$

T is called the period of the "difference frequency" between the two oscillators.

This means that at a series of equal time intervals, T, the two oscillators will be fully in step and at equally-spaced intervening intervals they will be out of step. For a short time on either side of the in step condition, their vibrations may be near enough to synchronization to allow an interaction. We will assume that this time interval is T/4 sec. For oscillators whose frequencies differ by 12.8 cm^{-1}, or 3.84×10^{11} Hz, the period of the difference frequency, T, is 2.60×10^{-12} sec and a quarter of this is 6.51×10^{-13} sec. Since this is *ex hypothesei* not enough time for an interaction between these two oscillators, it follows that the stimulus and the receptor must march in step for a time *greater* than about 10^{-12} sec. How much greater is not theoretically deducible, but an indication can be gained as follows. A molecule vibrating with a frequency of 200 cm^{-1} or 6×10^{12} Hz will execute fewer than ten oscillations in 10^{-12} sec. If it takes 100 such synchronized movements to effect an interaction, the interaction time will be of the order of 10^{-11} sec. We may take this as a rather rough minimum estimate of the minimum interaction time, based on the 12.8 cm^{-1} separation of the receptor frequencies.

When a gas molecule impinges upon a solid surface, it may rebound elastically or it may be adsorbed and reevaporated after a greater or lesser interval. The study of chemical reactions of gas molecules on the surface of solid catalysts generally suggests a finite residence time on the surface and that the molecule does not rebound elastically. Nevertheless, the elastic rebound represents the least possible time of contact between a stimulus and a receptor and it can be approximately estimated as follows.

We will assume that the molecule is perfectly elastic and the surface infinitely massive so that all the forces involved in the action are within the bouncing molecule. When the molecule makes its initial contact with the solid surface it is moving with a speed, v. As it begins to deform it is slowed down with a decelerating force, −kx, where k is the elastic force constant and x is the amount of deformation measured along the line of motion. Since (force) = (mass) × (acceleration), the deceleration, a, is −kx/m where m is the mass of the molecule. The speed of the molecule at any time during the impact is dx/dt and the rate of deceleration is

$$\frac{d}{dt}\left(\frac{dx}{dt}\right) = -\frac{kx}{m}$$

from which

$$\frac{dx}{dt} = -\frac{kxt}{m} + b$$

When t = 0, dx/dt = v, so that b = v. At the moment the molecule comes to rest before rebounding, dx/dt = 0, and

When t = 0, dx/dt = v, so that b = v

dx/dt = 0, and

$$\frac{kxt}{m} = v$$

so that

$$t = \frac{mv}{kx}$$

and the total in-and-out time is 2t.

The various terms on the right side of the equation can be estimated, and in the following calculation it will be assumed that the impinging molecule is benzonitrile, $C_6H_5{-}CN$, moving vertically towards the surface and with its long axis perpendicular to the surface.

At 300 K, the average translational velocity of a molecule with a gram molecular weight of 103 is given by the formula

$$v = 0.921\left(\frac{3RT}{M}\right)^{1/2}$$

R = 8.31×10^7 erg/mol, so that

$$v = 2.48 \times 10^4 \text{ cm/sec}$$

With a gram molecular weight of 103 and given Avogadro's number as 6.06×10^{23}, the kinetic energy, Q, of the molecule is 5.24×10^{-14} erg.

The elastic constant, k, can be estimated from the fact that the longitudinal, stretching vibration of benzonitrile is known from infrared spectroscopy to be 462 cm^{-1} or 1.39×10^{13} Hz. If we can regard the stretching vibration of this molecule as two groups of atoms joined by springs, we have

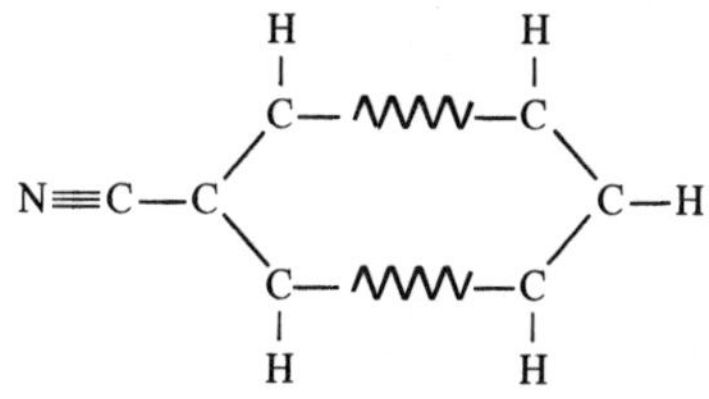

the frequency of the vibration is given by the formula

$$\text{frequency} = \frac{1}{2\pi}\left(\frac{k}{u}\right)^{1/2} = 462 \times 3 \times 10^{10}$$

where

$$u = \frac{m_1 \cdot m_2}{m_1 + m_2} = 4.00 \times 10^{-23} \text{ g}$$

u is called the "reduced mass".
From this,

$$k = 3.03 \times 10^5 \text{ dyn/cm}$$

At the moment of closest approach, when the molecule has momentarily come to rest, the energy is all potential, and if the elastic force constant is k and the displacement is x, then the potential energy is

$$Q = \int_0^x kxdx = \frac{kx^2}{2} = 5.24 \times 10^{-14}$$

from which

$$x = 5.88 \times 10^{-10} \text{ cm}$$

Substituting the various terms in the formula

$$t = \frac{mv}{kx}$$

gives

$$t = 2.37 \times 10^{-14} \text{ sec}$$

and

$$2t = 4.74 \times 10^{-14} \text{ sec}$$

which is the estimated time for an elastic rebound.

This calculation is necessarily approximate in that it ignores the effect of molecular rotation, as well as the possibility of an oblique descent of the molecule onto the surface, and the form of the oscillation was simplified by picturing the two ends of the molecule moving in and out as units. Yet it remains an estimate of the minimum possible time of contact between the stimulus and the receptor which can be compared with the estimate of the minimum interaction time which was based upon the 12.8 cm^{-1} separation of the receptor frequencies.

If this minimum contact time is gven a round value of 10^{-13} sec, and if the minimum interaction time is of the order of 10^{-11} sec, it suggests that there is very little attractive force to bind the stimulus to the receptor, and even less force to draw the stimulus towards the receptor when it is some distance away. Contact between stimulus and receptor is therefore primarily dependent upon the random processes of molecular diffusion, and while the interaction is not instantaneous the forces involved are only strong enough to arrest the rotation of the stimulus molecule but not strong enough to seriously perturb its vibrational frequencies. There may be some amount of perturbation with consequences to be discussed in Chapter 15, but any marked alteration in the stimulus frequencies would rule out the degree of correlation between odor and molecular vibration revealed by the peak number plots.

In the foregoing calculation, the molecule of benzonitrile was seen to present a specific profile to the surface by approaching with its long axis perpendicular to the surface. It could, of course, have presented other profiles with other orientations at the moment of impact. The particular orientation used for purposes of calculation was selected so that the elastic constant, k, could be given a numerical value. As explained in Chapter 8, molecular stereochemistry probably governs the probability of an interaction whose possibility is governed by molecular vibration.

Finally, occasional reports of radioactively tagged stimulus molecules being found chemically combined with something in the olfactory epithelium are probably less relevant to the processes of olfactory stimulation than to the accessory processes of cleanup or repair and maintenance of the receptor apparatus. As a fragile and dangerously exposed piece of highly sensitive apparatus, there must necessarily be a correspondingly diligent staff of housekeepers.

Chapter 12

WHAT KIND OF RECEPTOR?

The advantage of a good hypothesis consists essentially in deepening and broadening our knowledge of natural phenomena.

W. Nernst

Science rarely advances in a straight line. It moves, instead, in a zigzag path as ideas rebound off hard facts and often harsher criticisms. Yet in the process of growth, ideas which have been abandoned for good reasons have nevertheless had within them some essential something that still needs to be taken into account. Tracing the evolution of a pattern of ideas can therefore be of more than antiquarian interest.

In the evolution of theories of olfactory stimulation, it was at one time acceptable to speak of stimulus molecules nesting or settling themselves into receptor sites and somehow initiating nerve impulses before drifting or being somehow metabolized away. Here the essential fact was — and still is — olfactory specificity.

Later, opponents of the vibrational theory roundly asserted that there was no conceivable way a low-frequency, low-energy quantum of vibrational energy could bring about a much more energetic event, namely, the discharge of a nerve impulse. To answer this, it was suggested that the stimulus molecule triggers the nerve discharge by *de-exciting* a receptor which had previously become electronically excited by the metabolic processes inseparable from life itself.[1] Here, specificity was coupled with stimulation resulting from a *loss of energy* by the receptor rather than its acquisition of energy from the stimulus.

The next stage[2] focused on the idea of the stimulus somehow opening up the membrane of the receptor nerve cell so as to allow the passage of sodium and potassium ions, thereby bringing about a local depolarization of the sort commonly regarded as the first step in initiating a nerve impulse.

Given that the peak number plots identify the frequency-sensitivities of the receptors, their numerical values and their equal spacing are new and highly significant facts to be taken into account. To do this, it was supposed that the receptor site consists of two poles separated by a short gap and with an electric field between them which oscillates with a specific frequency so as to interact selectively with a molecule with a matching frequency that approaches the site. To account for the oscillating field between the poles, as well as for the equal spacing of the receptor frequencies, it was supposed that each pole was the end of a longitudinally vibrating molecule with the two poles coming in and out of phase so as to produce a difference frequency, the value of which would vary with the relative lengths of the two molecular chains so as to set the frequency sensitivity of the site.[3]

An engineering analysis by Professor C. A. Brockley of the possible values of the longitudinal vibrational frequencies of hydrocarbon or polysaccharide chains or polypeptide helices, revealed the hard fact that the longitudinal vibrations could not possibly generate the sort of unique receptor frequencies spelled out by the peak number plots.

Nevertheless, the concept of a bipolar receptor with an alternating electric field in the gap between the poles into which some part of the stimulus molecule would fit was an important idea, to be melded with the earlier ideas of specificity, de-excitation of the receptor, and the opening of the membrane to the passage of ions. For one thing, the need for the stimulus molecule to fit some part of itself into the gap separating the two poles of the receptor meant that even though a match in the vibration frequencies determines the *possibility* of an interaction, stereochemistry will largely determine its *probability*.[5]

What, then, can determine the frequency-specificity of a given type of receptor, and why are receptor frequencies evenly spaced?

It must be remembered that there is an electrical dipole associated with every covalent bond between unlike atoms, so that any relative movement of the atoms has an accompanying movement of the electric charges. Given that a molecule composed of N atoms can vibrate in 3N − 6 distinct ways, it is evident that as a mechanical structure a really complex molecule of the sort likely to be found in the membrane of a nerve cell could present an enormous number and variety of vibrational modes. It would be difficult for any one of them to dominate so as to establish a specific frequency to which a particular receptor is tuned.

This consideration holds as long as it is the strictly mechanical oscillations that generate and govern the electrical oscillations within the molecule. If, however, a way can be conceived for the roles to be reversed and for electrical effects to favor selectivity one of the mechanical vibrations, then the bipolar site tuned to a specific frequency becomes a potentially viable concept.

Such a process can be conceived as follows.

Consider a polypeptide chain which, for simplicity, will be taken as a chain of glycine units:

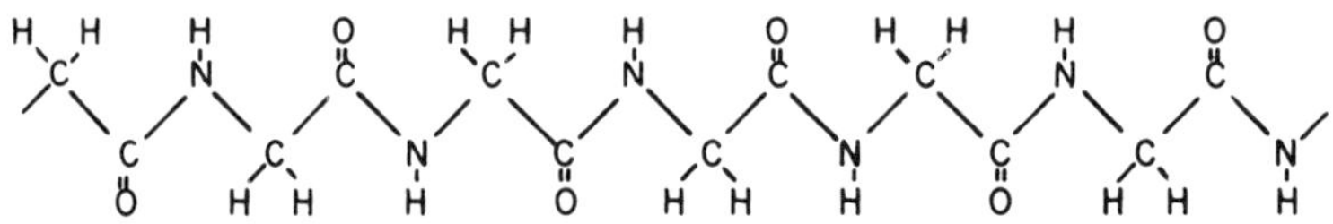

Spaced out along the chain there is a series of local dipoles. Suppose, now, that the polypeptide chain is bent around so as to bring the H_2N- group near to but not touching the carboxyl group and suppose that the proton from the carboxyl group attaches itself to the amino group as is known to happen in an ampholyte or zwitterion, represented in the simplest case by glycine:

$$H_3\overset{+}{N}—CH_2—C\overset{-}{O}_2$$

If the two ends of the loop are connected by a bridging group such as a $-CH_2-CH_2-CH_2-$ link, the structure will look something like this:

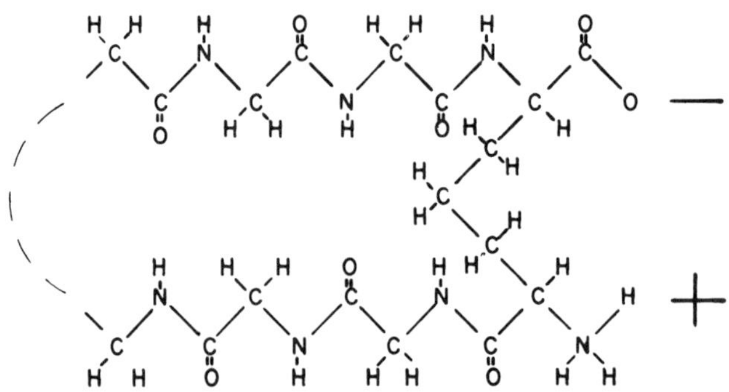

With this structure, the two polar ends are separated by a fixed distance, and if the space between them has insulating properties the two poles will function much like the plates of an electrical condenser. With some modicum of elasticity in the chain, the localized dipoles distributed along it will have some freedom to move to-and-fro. Thus, even though the chain has none of the characteristics of a metallic conductor, it can perhaps carry a current that alternates in a manner consistent with the stretching/contracting mechanical movements of its constituent, local dipoles. If we consider the polypeptide simply as a linear chain, it can be pictured as being twisted so as to place the local dipoles at an angle to the axis of the chain, in which case there would be self-induction when the current alternates. If the chain is then wound into a spiral or helical configuration there will be a coil within a coil, and the self inductance will be substantially augmented.

Given a bipolar receptor site with a capacity, C, together with a polypeptide chain with an inductance, L, the receptor molecule as pictured above will be capable of functioning as an electrical oscillating circuit with a frequency, F, given by,

$$F = \frac{1}{2\pi(LC)^{1/2}}$$

The capacity of a condenser depends upon the area of the plates, their separation, and the dielectric constant of the medium between them. Thus the value of C will be fixed by the conformation of the bipolar receptor site. For a given type of frequency-sensitive receptor, the inductance, L, will be fixed by the length of the polypeptide chain and the number of helical turns. Thus, each type of receptor has a unique frequency, F, corresponding to its specific values of L and C.

Receptors with different specific frequency-sensitivities are more likely to owe their specificity to a changed value of L than to an altered C, since L can vary with the length and conformation of the inductive loop more readily than C can change through a change in the geometry of the bipolar receptor. The even, 12.8 cm^{-1} spacing of the frequencies of the different receptor types is explicable because the inductance, L, will increase by increments as units are added to the chain, and with a long enough chain the increments will become roughly equal over a range of 40 to 50 steps.

Thus we have a bipolar receptor site on the surface of the nerve cell membrane with an electric field between the poles which does not reverse its direction but which does wax and wane with a specific frequency. For steric reasons, only some of the local dipoles in the stimulus can insert themselves into the gap between the poles of the receptor, somewhat as shown in Figure 1. This shows a model of the receptor site with a molecule of benzonitrile insinuating its −CN group into the gap. Bearing in mind that the stimulus molecule has a local dipole at each covalent bond joining dissimilar atoms, there will be a local dipole in the −CN group one end of which will be pulled or pushed to-and-fro by the oscillating field of the receptor. If the stimulus molecule has a normal mode of vibration with a matching frequency and if that mode has a component of motion parallel to the line joining the poles of the receptor, that normal mode can become excited by resonance. In the range of frequencies under consideration the transfer will be quantized, so that the receptor mechanism evidently straddles the boundary between classical and quantum mechanical physics. It may therefore lead to insights quite apart from the problem of olfactory stimulation.

The source of energy to excite or activate the molecular-scale oscillator to a level above that corresponding to the ambient temperature can come from chemical reactions necessary for the continued life of the organism. The process would be parallel to that taking place in a chemically pumped laser.

The de-excitation of the receptor-oscillator could alter and increase the ion-permeability of the membrane in various ways. It could, for example, desynchronize the movements of the receptor molecule and its neighboring proteinaceous or other molecules so as to open up the membrane to the passage of sodium and potassium ions. Alter-

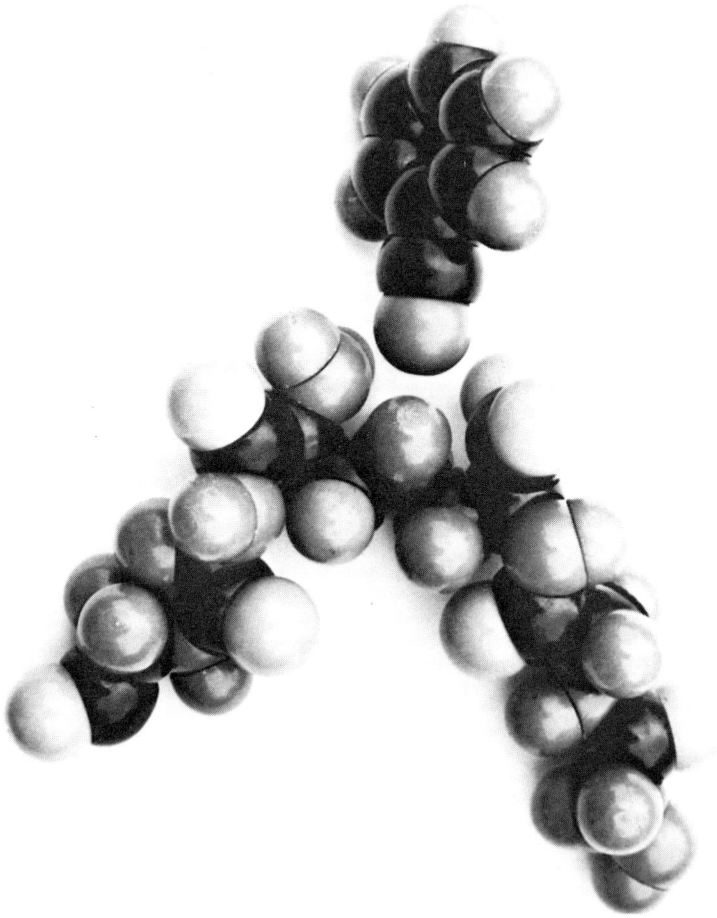

FIGURE 1. Model of a molecule of benzonitrile in the process of interacting with the suggested tuned electrical oscillator type of olfactory receptor. Only the out-of-plane vibrational modes of the benzonitrile will have an appreciable component parallel to the line joining the plates of the condenser so as to render them osmically active.

natively, it could open the membrane by altering the electric field in the vicinity of the bipolar receptor site.

Figure 1 throws an interesting light on the fact that the odor of benzonitrile is like bitter-almond. Its far infrared frequencies and their assignments are as follows:

Mode	Assignment
462 cm^{-1}	Longitudinal ring stretching, in-plane
401	Butterfly ring deformation, out-of-plane
381	Diagonal ring stretching, in-plane
172	Twisting ring deformation, out-of-plane
162	Side chain, C—N flexing

With the benzonitrile oriented with respect to the receptor dipole as shown in the figure, the in-plane vibrational modes would have little or no component of motion parallel to the line joining the poles of the receptor and so would not become excited. On the other hand, the out-of-plane modes would have such a component and so would be excitable. Figure 2, Chapter 6, shows that the 172 cm^{-1} frequency is consistent with a somewhat almond-like odor rather than a green-type odor which Figure 3, Chapter 6, suggests as likely to be generated by a frequency near 462 cm^{-1}.

REFERENCES

1. **Wright, R. H., Reid, C., and Evans, H. G. V.,** Odor and molecular vibration. III. A new theory of olfactory stimulation, *Chem. Ind.,* 973, 1956.
2. **Wright, R. H.,** Olfactory specificity, in Excerpta Medica Int. Congr. Ser. No. 206, Proc. 9th Int. Congr. Oto-Rhino-Laryngology, Mexico, D. F., August 10—14, 1969.
3. **Wright, R. H.,** Odor and molecular vibration: a possible membrane interaction mechanism, *Chem. Senses Flavor,* 62, 203, 1976.
4. **Wright, R. H.,** Odor and molecular vibration: neural coding of olfactory information, *J. Theor. Biol.,* 64, 473, 1977.
5. **Wright, R. H.,** Odor and molecular vibration: redundancy in the olfactory code, in *Odor Quality and Chemical Structure,* Moskowitz, H. R. and Warren, C. B., Eds., A. C. S. Symposium Series, 148, American Chemical Society, Washington, D.C., 1981.

Chapter 13

OBJECTION!

The habit of suspending judgment until a conclusion has been fully tested by varying the circumstances of the experiment and by repeated accurate measurement, is a valuable habit to acquire.

G. F. Fitzgerald

It has several times been remarked by critics of the vibrational theory of olfaction that replacing an atom or atoms in an odorous substance with isotopes of different mass should change the vibrational frequencies. This follows from the fact that the frequency of a harmonic oscillator depends upon the restoring force and the masses of the moving parts, so that although substituting a heavier isotope for a lighter one should have little or no effect on the forces, it should affect the frequency because of the change in mass. With the frequency changed (so runs the argument) the odor should change, and if it is not observed to change, so much the worse for the theory!

Since this is first of all a theoretical objection, let us look first at its theoretical foundation and then go on to examine the experimental evidence.

Quantitatively, for a simple harmonic oscillator, the frequency, ν, is given by the expression

$$\nu = \frac{1}{2\pi}\left(\frac{k}{u}\right)^{1/2}$$

where k is the restoring force constant and u is the reduced mass of the system, that is

$$u = \frac{m_1 m_2}{m_1 + m_2}$$

where m_1 and m_2 are the masses of the relatively moving parts.

Since the expression for the frequency involves the square root of the reduced mass, it is clear that relatively large changes in m_1 or m_2 will be needed before there is likely to be any important effect on the frequency. This is especially so with low-frequency vibrations which generally have a whole molecule character of the kind illustrated in Figure 1, Chapter 5. This means that, roughly speaking, one group of atoms is vibrating against another so that isotopically substituting some members of the group will have an even less effect than one might expect at first sight.

Isotopic substitution will without doubt affect the vibrational frequencies, but will the change be large enough to matter?

The question was put to its first real test by R. E. Doolittle and his colleagues with the U.S. Department of Agriculture.[2] A compound known as "cue-lure" or 4-(*p*-hydroxyphenyl)-2-butanone acetate, is a potent attractant for the male melon fly, *Dacus curcurbitae*. It has the structure

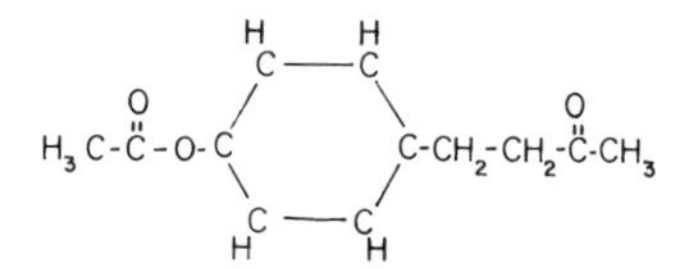

Table 1
FAR INFRARED FREQUENCIES OF CUE-LURE AND SOME DEUTERATED DERIVATIVES

Cue-lure	Derivative: A	B	C	D	E	Significant frequencies[a]
361	346	346	340	342	355	Neg.
328	323	323	312	316	300	310
—	—	—	—	—	—	298
225	225	224	219	224	224	236
199	195	194	192	192	199	200

[a] From Figure 1, Chapter 7.

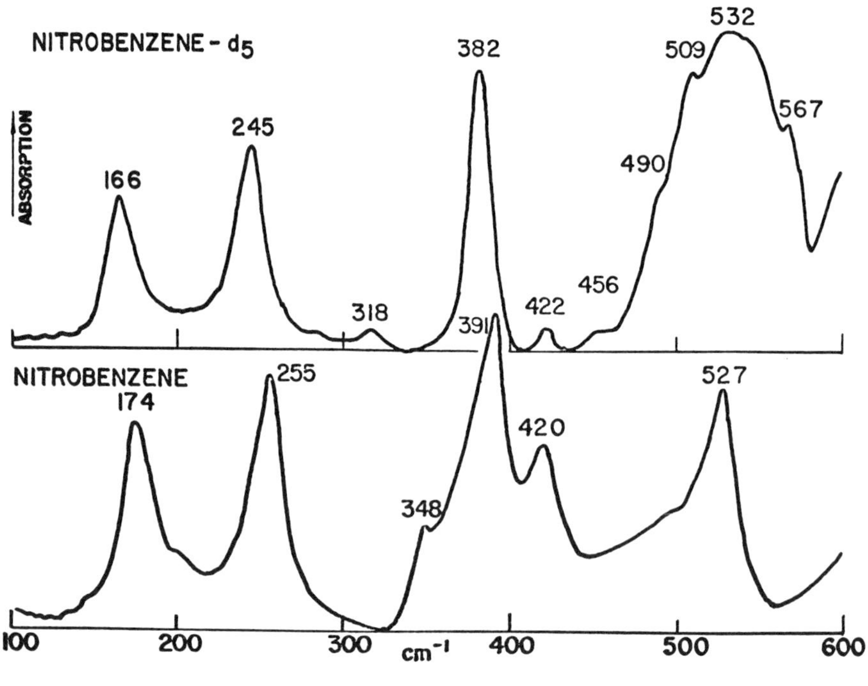

FIGURE 1. Far infrared spectra of ordinary nitrobenzene and fully deuterated nitrobenzene.

Hydrogen atoms at various places in the structure were replaced by deuterium atoms with mass 2 so as to load the structure with extra mass in a variety of different ways. There was no significant difference in the abilities of the various preparations to attract the flies.

However, the far infrared spectra showed that in conformity with the theory outlined above, the overall frequency-patterns were not changed enough to make a change in the biological response at all probable. Table 1 shows the vibrational frequencies of the variously deuterated cue-lures, as well as those of the ordinary, undeuterated compound. From Figure 1, Chapter 7, not all these frequencies need to be biologically significant, and the ones that do seem to be significant are the ones least affected by the deuteration.

The next experiment of this kind was made by R. J. Barker and others at the U.S. Government's Bee Research Laboratory.[1] Worker bees were trained to associate the smell of ordinary nitrobenzene, $C_6H_5NO_2$, with a source of sugar syrup, and then given the choice of ordinary nitrobenzene and fully deuterated nitrobenzene, $C_6D_5NO_2$. The bees made no distinction between the two, showing that their perception of the odors was not affected by the deuteration.

Dr. Barker was good enough to send me the deuterated nitrobenzene left over from his experiments so that I could compare its vibration frequencies with those of ordinary nitrobenzene. The far infrared spectra are shown in Figure 1, and once again the frequencies are changed but not enough to make one anticipate a serious change in the odor.

It is interesting to see how well the formula for a simple harmonic oscillator given above applies to the real situation.

The stretching frequency shown at the middle of Figure 1, Chapter 5 has a frequency near 390 wave numbers in ordinary nitrobenzene. To a first approximation it can be pictured as an upper block of two −CH− groups plus a $-CNO_2-$ group vibrating against a lower block of three −CH− groups. The masses of the two groups are, respectively, 84 and 39, giving a reduced mass for ordinary nitrobenzene of approximately 26.6 which becomes 28.2 in the deuterated compound. The difference in their square roots is 0.15 or about 2.9%. The observed shift in frequency is 9 wave numbers, or about 2.3%.

There is one case where deuteration might be expected to affect the odor. Ordinary naphthalene has far infrared frequencies at 363 and 183 wave numbers which are shifted to 339 and 169 by complete deuteration. Six observers were given two samples of ordinary naphthalene and one of the deuterated compound and asked which one of the three was in any way different from the other two. Four of the six correctly identified the deuterated sample, one reported "no difference" and one got it wrong. Bearing in mind the difficulties in using human judges which have already been mentioned (and will be mentioned again!) this can be considered evidence for an effect of isotopic substitution in this particular case.

Summing it up, the decision must be "objection overruled".

REFERENCES

1. **Barker, R. J., Berdel, R. L., and Waller, G. D.,** The molecular basis of scent discrimination: response to nitrobenzene-d_5 of honey bees (*Apis mellifera* L.) conditioned with nitrobenzene, *Experientia,* 29, 418, 1973.
2. **Doolittle, R. E., Beroza, M., Keiser, I., and Schneider, E. L.,** Deuteration of the melon fly attractant, "cue-lure" and its effect on olfactory response and infrared absorption, *J. Insect Physiol.,* 14, 1697, 1968.
3. **Wright, R. H.,** Odor and molecular vibration: response to nitrobenzene-d_5 of honey bees (*Apis mellifera* L.) conditioned with nitrobenzene, *Experientia,* 31, 530, 1975.

Chapter 14

OBJECTION — II

I am defending the thesis tonight that an experiment should never be trusted until it has been checked by a good theory. Many people take the opposite view; but I intend to show that they are wrong.

W. D. Bancroft

Chemical compounds are known to exist in which the molecules contain the same number and the same kinds of atoms, but which nevertheless have different properties because of a difference in the ways the atoms are connected together. A very simple example would be the two substances with the general formula, $C_2H_4Cl_2$:

$Cl—CH_2—CH_2—Cl$	and	$CH_3—CHCl_2$
symmetrical dichlorethane		unsymmetrical dichlorethane

Compounds with the same molecular formula but different structures and properties are called "structural isomers".

A more subtle form of isomerism arises when the same number and kind of atoms are connected *in the same ways* but in such a fashion that one cannot be superimposed on the other because they are mirror images, like a right and a left hand. A right hand which has the same structure and dimensions as the left one will nevertheless not go into a left-handed glove. Figure 1 shows models of two such mirror-image molecules, the two isomeric carvones. This phenomenon of handedness, as of an object and its image in a mirror, is called "chirality", and the two forms are said to be "enantiomers" or to constitute an "enantiomeric pair". Chirality arises whenever the three-dimensional structure has no plane of symmetry, that is, when there is no way the structure can be sliced in two and the whole then restored by placing either half against a mirror.

Classically, enantiomeric pairs are molecular species with the same structural formulas *except* that they are mirror images when looked at in three-dimensional space. They have identical chemical properties *except* when interacting with something that is itself asymmetric. They have identical physical properties *except* when they are interacting with electromagnetic radiation, and so are often called "optical isomers".

Since the masses and elastic forces in a chiral pair are necessarily identical, it has generally been assumed that the vibrational properties of enantiomers will be identical also, and if this is so the natural conclusion would seem to be that they should have identical odors. This has led more than one critic of the vibrational theory to comment that if the existence of enantiomers with different odors could be demonstrated it would "spell death to any theory of olfaction which looks for a correlation between odor quality and infrared spectra."[1] In this chapter we will look at the evidence and in the next we will appeal the death sentence.

Until quite recently there was no reliable evidence that optical isomers could, in fact, have different odors. For one thing, the odor threshold concentrations of most fragrant chemicals are so far below the usual techniques of chemical analysis that observed differences could be attributed to traces of impurity. Moreover, in most cases with really carefully purified samples there was (and is) no perceptible difference in

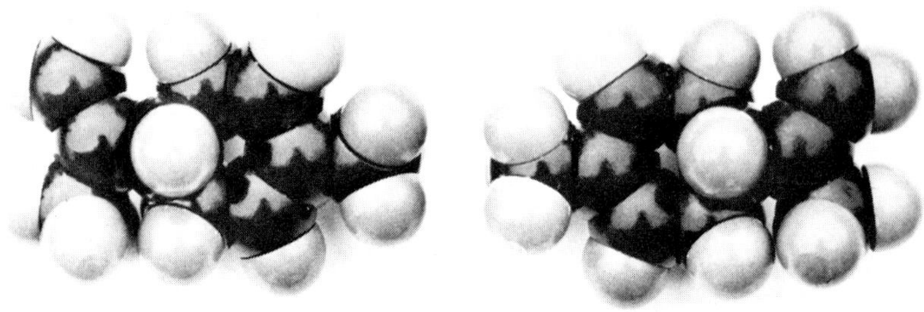

FIGURE 1. Models of the enantiomeric carvone molecules which are related as an object and its mirror image. The structural formulas cannot be made to superimpose without lifting one out of the plane of the paper and turning it over.

either the human or the insect response. Thus, for example, M. Jacobson[4] tested the enantiomeric forms of the sex-attractant of the gypsy moth, *Porthetria dispar,* and found that the males were equally drawn to minute amounts of each.

Since about 1970, however, a number of reports have appeared to indicate that some — but by no means all — enantiomeric pairs do have recognizably different odors when smelled by some — but by no means all — human observers.

By way of example, T. J. Leitereg and his colleagues[5] found that of the two carvones shown in Figure 1, one smelled like caraway and the other like spearmint. Their samples were prepared and purified as carefully as possible but even so the possibility remained that a minor impurity could have been responsible for the observed dissimilarity. Nevertheless, "the odor characteristics of the natural and synthetic carvones were evaluated by an odor panel of 21 to 26 persons who had extensive experience and proven reliability in odor judgments..." The ratio of correct to incorrect judgments in four kinds of experiments was 21:4, 17:5, 20:5 and 17:5. These are statistically significant but hardly indicative of a very marked difference in the odors. An independent study of the same carvones by G. F. Russell and J. I. Hills[9] using a triangle test (two samples the same and the observer asked to find the odd one) never got better than 92% correct judgments.

The verbal descriptions of trained experts in perfumery are perhaps more revealing of the sort of differences that may exist. G. Ohloff[8] has reported on the odorous properties of several enantiomeric pairs with the results shown in Table 1. Once again, the differences appear to be real but on the whole rather slight.

What has never been demonstrated using human observers is a case where one member of a pair has an odor and the other has no odor at all.

As before, when human evaluations are at best fuzzy, it is helpful to call in insects to serve as judges. Y. Lensky and M. S. Blum[7] trained honeybee workers to go to one of the carvones, using the reflex extension of the proboscis as a signal that the odor was being perceived. In 44 tests with bees conditioned to one of the carvones, 41 positive responses were recorded with that one and only 3 with the mirror-image carvone. Again a difference, but not an absolute one.

A very curious instance has recently been reported by J. H. Borden and his coworkers.[2] The aggregation pheromone of the beetle, *Gnathotrichus sulcatus,* has been isolated and chemically identified as 6-methyl-5-hepten-2-ol, or sulcatol. Its molecules have no plane of symmetry and so can exist in enantiomeric forms. In both laboratory and field bioassays, the insects responded only when both isomers were present and the pure enantiomers were each inactive. The natural secretion is a 65:35 mixture of the two.

Thus the overall situation as regards odor differences in enantiomeric pairs is best summed up in the words of L. Friedman and J. G. Miller[3] who unambiguously con-

Table 1
OLFACTORY PROPERTIES OF ENANTIOMERIC FRAGRANCES

Enantiomers	Olfactory properties
(+)-R-4-Methyl caproic acid	Caproic type odor
(−)-S-4-Methyl caproic acid	Same shade but more fatty and more powerful
(+)-3S-Citronellol	Citronella oil type odor
(−)-3S-Citronellol	Geranium type odor, more powerful
(+)-*cis*-Rose oxide	Green, geranium type odor: *cis*-derivatives more powerful than the corresponding trans; (+)-compounds sweeter and less herbal than (−)-; (−)-*cis*-rose oxide most powerful.
(−)-*cis*-Rose oxide	
(−)-*trans*-Rose oxide	
(+)-*trans*-Rose oxide	
(−)-3R-Linalool	Woody lavender-like odor more powerful
(+)-3S-Linalool	Sweety shade of lavender-like odor, more reminiscent of petit grain oil
(−)-S-7-Hydroxy citronellal	Reminiscent of the lily of the valley with a fresh green and minty shade
(+)-R-7-Hydroxy citronellal	Strong sweet lily of the valley-like odor, more powerful

firmed the odor difference between the two carvones: "Lest the impression be created that all enantiomers possess different odors, it should be stated that there are many examples to the contrary. However, it can be concluded that enantiomers *can* have different odors..."

This, then, is the case for the prosecution. Let us now turn to the case for the defense.

REFERENCES

1. **Anon.**, Smells left and right, *Nature (London)*, 232, 372, 1971.
2. **Borden, J. H., Chong, L., McLean, J. A., Slessor, K. N., and Mori, K.**, *Gnathotrichus sulcatus*: synergistic response to enantiomers of the aggregation pheromone, sulcatol, *Science*, 192, 894, 1976.
3. **Friedman, L. and Miller, J. G.**, Odor incongruity and chirality, *Science*, 172, 1044, 1971.
4. **Jacobson, M.**, Insect sex attractants. III. Optical resolution of dl-10-aceto-*cis*-7-hexadecen-1-ol, *J. Org. Chem.*, 27, 2670, 1962.
5. **Leitereg, T. J., Guadagni, D. G., Harris, J., Mon, T. R., and Teranishi, R.**, Evidence for the difference between the odors of the optical isomers of (+)- and (−)-Carvone, *Nature (London)*, 230, 455, 1971.
6. **Leitereg, T. J., Guadagni, D. G., Harris, J., Mon, T. R., and Teranishi, R.**, Chemical and sensory data supporting the difference between the odors of the enantiomeric carvones, *J. Agric. Food Chem.*, 19, 785, 1971.
7. **Lensky, Y. and Blum, M. S.**, Chirality of insect chemoreceptors, *Life Sci.*, 14, 2045, 1974.
8. **Ohloff, G.**, Odorous properties of enantiomeric compounds, in *Olfaction and Taste 4*, Wissenschaftliche Verlag, Stuttgart, 1972, 156.
9. **Russell, G. F. and Hills, J. I.**, Odor differences between enantiomeric isomers, *Science*, 172, 1034, 1971.
10. **Theimer, E. T. and McDaniel, M. R.**, Odor and optical activity, *J. Soc. Cosmet. Chem.*, 22, 15, 1971.

Chapter 15

FOR THE DEFENSE!

When a fact appears to be opposed to a whole train of deductions, it invariably proves to be capable of bearing some other interpretation.

Sherlock Holmes

Inasmuch as enantiomeric (mirror image) molecules are presumed to have identical molecular vibrational frequencies, any difference in their odors has been made ground for categorically denying the possibility that odor or pheromone activity can be in any way related to molecular vibration.

Is this denial valid?

There can be no doubt that in the absence of external, perturbing forces, enantiomers display identical frequency-patterns, but it cannot be assumed that this will still be true when such forces begin to operate. In Chapter 9 we saw how the vibrational frequencies of naphthalene are to some extent modified when it is dissolved in benzene. Evidently, close association with the unrelated solvent molecules can alter the vibration patterns, presumably because interacting force-fields modify the interatomic forces within the molecule.* Neither benzene nor naphthalene is chiral but their example serves to illustrate the basic fact.

The situation when a nonchiral solute is dissolved in a chiral solvent has been examined by Hayward and Totty with results that are extremely significant.[2]

Briefly, this is the situation. Consider, first, the phenomenon of optical activity which is the distinctive property of optical isomers. Classically, it is the ability of each enantiomer to rotate the plane of polarization of plane-polarized light in a clockwise or counterclockwise direction. But plane-polarized light can be regarded as the resultant of two circular polarizations of equal amplitude and phase, but opposite direction. The rotation of the plane of polarization by an optically active compound is a result of a difference in the way the molecule interacts with the two circular polarizations. This has been verified experimentally when it was shown that an optically active substance (such as one of the two carvones mentioned in the last chapter) will, in fact, show different refractive indices and absorbances for light or infrared radiation which is circularly polarized in one direction or the other. The phenomena are well-known and are called "optical rotatory dispersion" and "circular dichroism", respectively. See Figure 1.

Compounds which are optically inactive by the usual test with a polarimeter using plane-polarized light will exhibit circular dichroism *if placed in an asymmetric environment.*

The dichroism is induced by the solvent on the solute and without actual binding: it can take place while their respective molecules are separated by several Van der Waals radii. The effect is no doubt due to the fact that the intermolecular forces fall off as an inverse fourth or higher power of the distance so that the force field surrounding a molecule has a very steep gradient. As a stimulus molecule moves in, its nearer side will be pulled out of shape more than its far side. Like all living structures, a receptor site is almost certainly asymmetric and so its force field is asymmetric, and the resulting deformation of the approaching molecule must be asymmetric. Furthermore, the induced asymmetry will not be the same for enantiomeric pairs, and so, even though their vibrational frequencies may be the same in isolated space, they are no longer

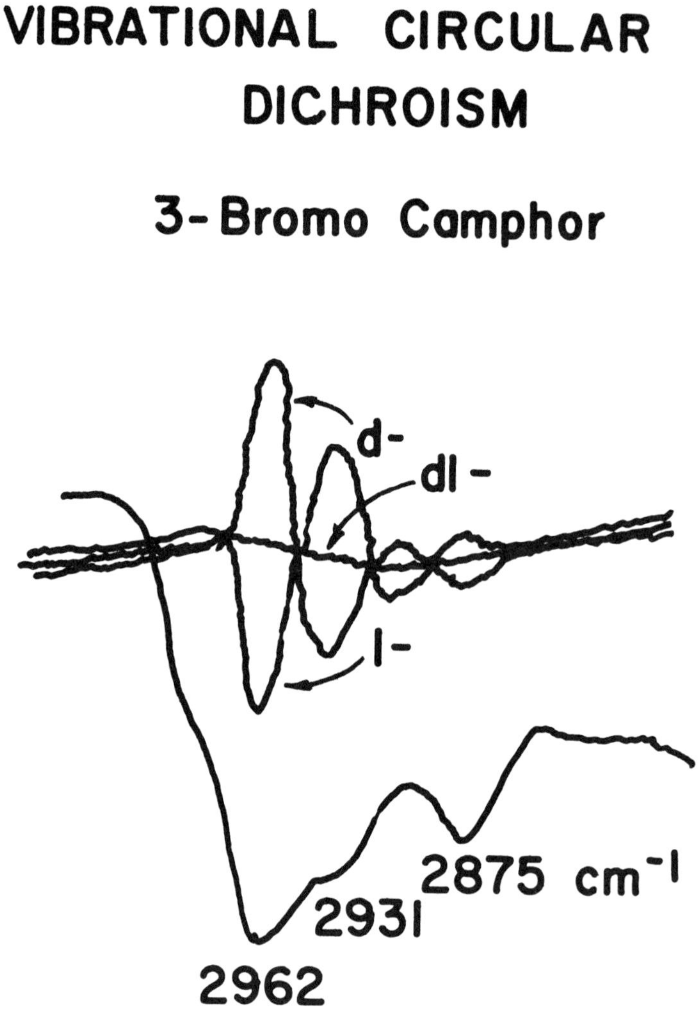

FIGURE 1. Vibrational circular dichroism of camphor. Curve A shows the near infrared absorption of camphor recorded with unpolarized radiation. Curve B shows absorption by l-camphor and Curve C shows transmission by d-camphor when using circularly polarized radiation. With racemic, or dl-camphor and circularly polarized radiation, the two effects cancel, to give Curve D. (From Nafie, L. A., Keiderling, T. A., and Stephens, P. J., *J. Am. Chem. Soc.*, 98, 2715, 1976. With permission.)

precisely the same in the near vicinity of a receptor. The effect will naturally be greater with a stimulus molecule with a rather loose, flexible structure than in one that is relatively rigid.

In short, in a chiral environment (such as exists in the immediate vicinity of a receptor site) asymmetry is induced in normally symmetric substances, and in molecules which are already chiral, the added twist will not be the same for the dextro-isomer as for the laevo-isomer when placed in an environment of some given chirality. They will no longer be mirror images, and *their vibration patterns will no longer be identical!*

Thus, the facts are clear. Their relevance to the vibrational theory of odor is simply this: given that the living molecular receptor-sites are almost certainly chiral, the frequency-pattern of a dextro-isomer can be in some degree different from that of its mirror image, laevo-isomer *at the moment of interaction.*

Whether the difference will be great enough to be significant in altering the olfactory response will depend upon factors that will differ from case to case. A receptor sensitive to a given stimulus frequency may respond differently to different molecules even though they may have numerically identical frequencies if the frequencies originate in vibrational modes of different symmetry. In the same way, inasmuch as the various vibrational modes of a given stimulus have different symmetries, the presence or absence of the chiral effect in the stimulus-receptor interaction will vary as between the several normal modes of the stimulus.

It is clear, then, that occasional reports of qualitative differences in the odors or insect-attracting properties of enantiomers are not a fatal objection to the vibrational theory. Furthermore, cases where an insect will respond only when both enantiomers are present can be regarded in precisely the same way as cases where the natural pheromone is a mixture of chemicals rather than a single chemical individual.

If the vibrational attributes of a molecule are subject to modification when in the near vicinity of an olfactory receptor site, the question may well be asked whether the ordinary, far infrared spectra can be made the basis of a valid theory of olfaction.

Evidently, the spectra recorded in the usual way with a solvent like benzene, or even when recorded in the vapor phase, are by no means ideal. Even so, such spectra are for the time being the best and, indeed, the only means of securing a useful insight into the process by which an odorous molecule is able to impress its unique odorous character upon the organism. The record of predictive successes of the theory and the range of phenomena it incorporates and accommodates, testify to the fact that the kind of frequency-perturbations we are considering are of no more than secondary importance. This is why, in most cases, there is so little difference in the human response to enantiomeric stimuli. Insects appear to discriminate more sharply in some cases, especially those involving species-specific sex pheromones. Probably this is because the perception of these is registered through specialist receptors, to be discussed later, in Chapter 21.

Thus, to conclude, the far infrared, low-frequency oscillations of molecules can be the molecular attributes which underlie olfactory specificity, but their ability to interact with chiral receptor molecules will depend upon the natural or induced chirality of the stimuli which will, in turn, depend upon the symmetry of each of the vibrational modes.

* The interaction will, of course, be mutual and inseparable from any process of chemorecognition, whether it be an olfactory perception, an immunological response, or, indeed, almost any of the vital processes of a living organism. Olfaction is a type of chemorecognition or chemoreception which transmits information to the CNS. In other types of chemorecognition, such as antibody-antigen reactions, the end result will naturally be different even though the chiral aspects may have much in common. See L. D. Hayward, *Nature*, 267, 554, 1977.

REFERENCES

1. **Craig, D. P. and Mellor, D. P.,** Discriminating interactions between chiral molecules, *Top. Curr. Chem.*, 63, 1, 1976.
2. **Hayward, L. D. and Totty, R. N.,** Optical activity of symmetric compounds in chiral media. I. Induced circular dichroism of unbound substrates, *Can. J. Chem.*, 49, 624, 1971.

3. **Hayward, L. D.,** A symmetry rule and mechanism for optical activity induced in achiral ketones by chiral solvents, *Chem. Phys. Lett.,* 33, 53, 1975.
4. **Hayward, L. D.,** A new theory of olfaction based on dispersion-induced optical activity, *Nature (London),* 267, 554, 1977.
5. **Nafie, L. A., Keiderling, T. A., and Stephens, P. J.,** Vibrational circular dichroism, *J. Am. Chem. Soc.,* 98, 2715, 1976.
6. **Wright, R. H.,** Odor and molecular vibration: optical isomers, *Chem. Senses and Flavor,* 3, 35, 1978.

Chapter 16

STIMULUS VS. RESPONSE

"It's the oldest rule in the book," said the King. "Then it ought to be Number One." said Alice.

Lewis Carroll

Psychophysics is the name given to a branch of science which seeks to relate as quantitatively as possible the strength or intensity of a sensation with the measured magnitude of the stimulus that evokes it. How, for example, does our estimate of the weight of an object or the brightness of a light relate to the measured mass of the object or the candle power of the light?

The first attempt to answer this sort of question quantitatively was made by E. H. Weber in 1834 when he invented the method of "Just Noticeable Differences" (JND). In this method, you are asked whether two weights, or lights, or sounds, or smells are just noticeably different, and when you have decided that there is such a barely perceptible difference, the actual difference is measured in the appropriate physical units and compared with that of the lighter or weaker of the two. When this is done, it turns out that the increment needed to produce a JND is related directly to the stimulus it is added to. Thus, for example, if you are listening to one or two voices you will easily notice when another joins in, but if you are listening to a chorus of 100 voices, adding one more will probably not be noticed. Or, again, if your take-home pay is $10 a day and is suddenly increased by $10, you will feel the difference more than you would from a $10 addition to a base rate of $100 a day.

Mathematically, Weber's Law can be written

$$\frac{\Delta C}{C} = \text{a constant for one JND}$$

It appears to apply to most estimates of the strength of most sensations.

Fechner carried Weber's work a stage farther by pointing out that if dC/C is constant for a JND in the sensation, then

$$S = a\mathrm{Log}C + b$$

where S is the estimated feel of the stimulus, C is the objectively measured strength of the stimulus, and a and b are constants. At or just below the threshold of perception,

$$S = a\mathrm{Log}\,\frac{C}{C_0}$$

This is called the Fechner-Weber (or Weber-Fechner) Law and is a fundamental law of Psychophysics (a name coined by Fechner to designate the new science he was founding).

There have been several researches to establish whether it applies to smell perception. One of the earliest and still one of the most thorough was carried out by E. A. McC. Gamble.[2] Using the method of JND, she found that despite experimental difficulties from fatigue of the organs of smell, personal idiosyncracies and so on, the concentration of odorous material has to be increased by about one third for the effect to be just noticeable. Others who have used the same method have reported somewhat higher

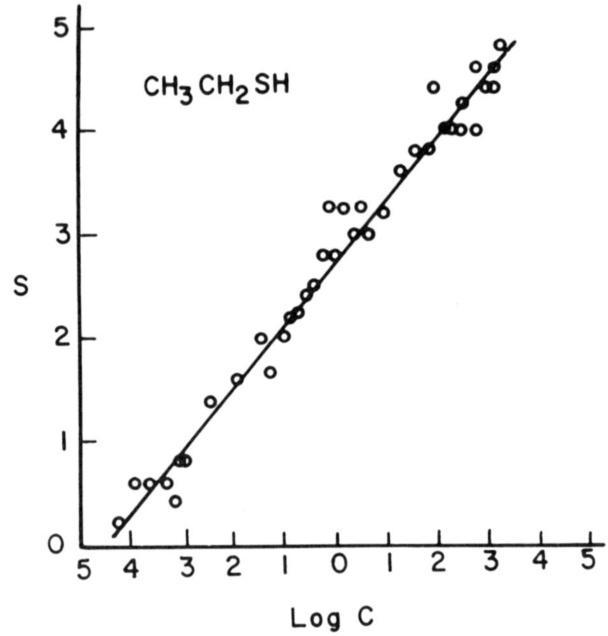

FIGURE 1. The Fechner-Weber plot of Fieldner's data for the perceived odor intensity of ethyl mercaptan as a function of concentration. The sensation, S, is related to the stimulus concentration, C, by the formula S = 0.57LogC + 2.7, indicating that the concentration-change necessary for a Just Noticeable Difference is about 57%.

values; in at least one case a 58% increase in the stimulus was needed to produce a JND.

It is not surprising that there should be this kind of disparity in the results. Biologically, the nose is not called upon to make precise estimates of odor strength. What it does, and as we showed in Chapter 2 it does supremely well, is to make precise discriminations between odors. It has been fashionable in scientific circles to pursue quantitative precision regardless of whether the thing being measured is in fact precisely measurable. Some of the emphasis which over the years has been given to precise measurement of JND in *intensity* might better have been placed on precise measurements of JNDs in odor *quality.* But let us return to the Fechner-Weber Law.

Where Gamble and others used the method of JNDs to verify this Law, A. C. Fieldner and his colleagues[1] with the U.S. Bureau of Mines applied the Law in its integrated form

$$S = a\text{Log}C + b$$

They were interested in the possibility of sending a danger signal to coal miners by adding a powerful smell to the ventilation system. For this, they proposed to use ethyl mercaptan, a substance with a powerful and rather distinctively unpleasant smell. A panel of six observers was used to estimate the strength of the smell on a 0 to 5 scale which they were supposed to set up in their minds in the following way.

0 or no odor requires no amplification; degree 1 is the threshold odor, the just perceptible odor. Consider now the opposite end of the scale: degree 5 or very strong is a most intense odor; it must be perceived,

Table 1
Odor Strength vs. Concentration: JNDs Constant

Numbers of JNDs	Concentration	Description
	1.0	
1	1.5	
2	2.2	
3	3.4	Just perceptible
4	5.1	
5	7.6	
6	11.4	
7	17.1	
8	25.6	
9	38.4	Distinct
10	57.7	
11	86.5	
12	130.0	
13	195.0	
14	292.0	
15	438.0	Modcrate
16	657.0	
17	986.0	
18	1,479.0	
19	2,218.0	
20	3,330.0	
21	4,990.0	Strong
22	7,480.0	
23	11,230.0	
24	16,850.0	
25	25,300.0	
26	37,900.0	
27	56,800.0	Overpowering
28	85,300.0	
29	128,000.0	
30	192,000.0	

however, only as an odor and aside from any other physiological effects caused by the chemical, such as irritation, nausea, or otherwise; degree 3 or easily noticeable is the medium odor midway in intensity between degrees 1 and 5; degree 2 or faint is conceived as midway between degrees 1 and 3; similarly degree 4 or strong is conceived as midway between degrees 3 and 5. With this scale of 0 and 5 equal degrees the observers recorded their impressions of the odor intensities. They did not always agree closely, but it was found that by employing a number of observers (six generally) and taking the averages of their observations, very satisfactory measurements of the odor intensities could be taken.[1]

One of their observers was not very discriminating and gave nearly all the mixtures the same value, 2, and as he was used in less than half the experiments I have replotted their results leaving him out to produce the graph shown in Figure 1. Each circle represents the average of five or occasionally four separate estimates of the smell strength. Despite the extremely subjective method of giving quantitative, numerical ratings to the sensations, the straight-line relation called for by the formula is very well-shown. What is even more striking is the spread between the strongest and the weakest concentrations of ethyl mercaptan: the range from $\text{LogC} = -4$ to $\text{LogC} = +3$ corresponds to a situation in which the strongest stimulus is ten million times more concentrated than the weakest.

Table 1 shows a series of numbers starting with 1 and increasing them each time by a factor of 1.5. This corresponds to a Weber Ratio, $\Delta C/C$, of ½ or, in words, to produce a JND we must increase any stimulus by 50%. This shows how the nose achieves the amazing dynamic range illustrated in Figure 1, because between the strongest and weakest sensations there are no more than 30 or 40 JND. It is important to realize, however, that at the lower concentrations, relatively small absolute increments in the stimulus are noticeable.

The words shown at the right of the table are descriptive terms often used to register odor levels. They are less ambiguous than such indefinite comparatives as fairly weak or rather strong, and more useful for practical purposes than expressions like perfectly disgusting. The value of this kind of numerical rating will be explained more fully in Chapter 34.

For the moment, it appears that the basic Fechner-Weber Law is well supported by two independent experimental techniques.

REFERENCES

1. **Fieldner, A. C., Sayers, R. R., Yant, W. P., Katz, S. H., Shohan, J. B., and Leitch, R. D.,** Warning Agents for Fuel Gases, Monograph No. 4, U.S. Bureau of Mines, Department of Interior, Washington, D.C., 1931.
2. **McC. Gamble, E. A.,** Applicability of Weber's Law to smell, *Am. J. Psychol.*, 10, 82, 1898.

Chapter 17

STIMULUS VS. RESPONSE — II

"Curiouser and curiouser!" cried Alice.

Lewis Carroll

In recent years the Fechner-Weber Law has been in some degree superseded as a result of work by S. S. Stevens[3] and others. For various reasons, Stevens believed that the best way to attach a number to a sensation is to assign a definite value to one sensation and then ask the observers to say whether some other sensation is one half, or some other fraction of it, or whether it is two or three or whatever number of times stronger than the standard. This is different from what Fieldner did, because he asked his observers to make a mental picture of a scale marked off in zero to five equal divisions and decide whereabouts on the scale they thought the smell should come.

One reason for Stevens proposing his Ratio Estimation method was that the uncertainty in the ratio could be compared with the size of the ratio, just as the just noticeable change in the stimulus could be related to the size of the stimulus.

To see what this means, let us go back to the Fechner-Weber Relation. There we work with one fixed stimulus and sensation, S, and vary the concentration, C, of a second stimulus to see how much the variation, ΔC, from the first stimulus must be to make a Just Noticeable Difference (JND). With the Stevens method of Ratio Estimation, we do the opposite: we present two stimuli in a fixed ratio and look to see how much the individual estimates vary. Then, just as Weber compared ΔC with C, so Stevens compared ΔS with S.

Because two different shades of grey, for example, retain the same relative brightness in a dim light as in a bright one, Stevens put

$$\frac{\Delta S}{S} = n \frac{\Delta C}{C}$$

where S is the sensation measured by the method of Ratio Estimation, C is the stimulus measured in appropriate physical units, and n is a constant of proportionality. Then the ordinary operations of mathematics show that

$$S = \left(\frac{C}{C_0}\right)^n$$

or, alternatively

$$\text{LogS} = \text{nLogC} - \text{nLogC}_0$$

where C_0 is the threshold stimulus.

To verify this formulation, we measure S by the method of Ratio Estimation and plot LogS against LogC and look for a straight-line relation. This has been done by F. N. Jones,[1,2] for example, using solutions of benzene in an odorless mineral oil. He assigned a strength of 10 to the middle solution of the series and asked 36 observers to relate all the solutions in relation to the standard, and averaged the results. His data are plotted in Figure 1 and show a reasonably straight line.

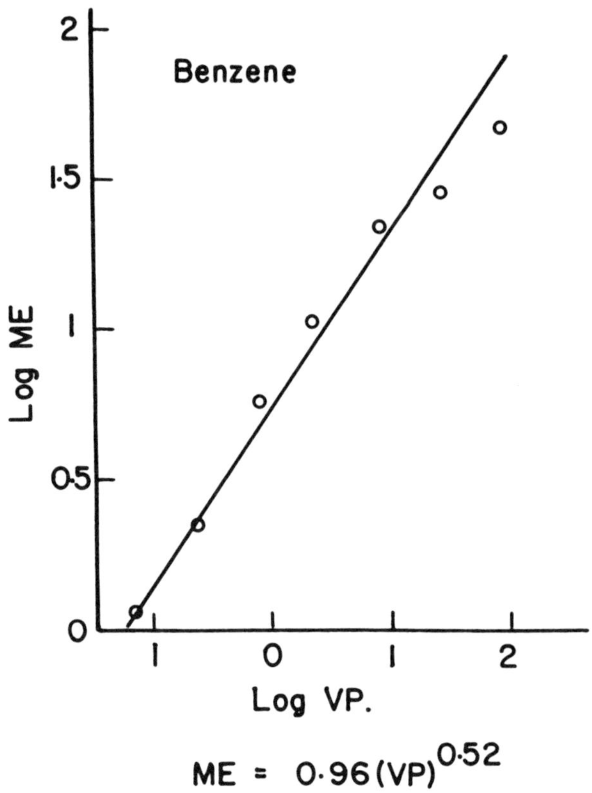

FIGURE 1. The Stevens Plot of F. N. Jones MEs (Magnitude Estimates) of the relative odor strengths of solutions of benzene of varying concentration, as compared with a standard. The Log-Log plot of the results gives an approximately straight line.

The same thing has been done with several other odorous substances but none of them had very low thresholds so that he did not cover the millionfold range of concentrations used by Fieldner. The Stevens Ratio Estimation Method was, however, validated.

The difference between the Fechner-Weber plot and the Stevens plot is that in the first, one plots S against LogC, and in the second it is LogS against LogC. If one plots Fieldner's results by the Stevens method one gets a strongly curved line because the two methods of plotting are different. This reflects the fact that the experimenters were asked two entirely different questions. Stevens Law and the Fechner-Weber Law are equally valid, equally empirical generalizations from experiment, and together they are the basic laws of psychophysics.

The special advantage of the Stevens approach is that it opens up a new and interesting line of enquiry.

It is fairly easy for an observer looking at two white lights to judge whether they are, in his opinion, equally bright or whether one is, say, twice as bright as the other. He is asked a very different question if he is required to compare the brightness of a white light and a green one. Nevertheless, the experiments show that quite respectably reproducible estimates can be made in this way. What is even more interesting is the fact that this kind of comparison can be made between such unrelated sensations as the loudness of a sound and the intensity of a vibration felt with the fingers. This is called a "cross-modality match" and many kinds of cross-modality comparisons have been tried with success. Some, indeed, have been carried around a circle where A is compared with B, B with C, and then C with A.

Mathematically, this system of round-robin ratio estimating is formulated as follows. If we have two sensations, S_1 and S_2, (they may be a light and a sound, for example) which are at the same level above threshold, then

$$S_1 = \left(\frac{x_1}{x_0}\right)^n$$

$$S_2 = \left(\frac{y_1}{y_0}\right)^m$$

and if $S_1 = S_2$ as judged by a panel of observers

$$\left(\frac{x_1}{x_0}\right)^n = \left(\frac{y_1}{y_0}\right)^m$$

or

$$\text{Log}\ \frac{x_1}{x_0} = \frac{m}{n}\ \text{Log}\ \frac{y_1}{y_0}$$

Given that the two Ss are the same, this means that the stimulus ratios should be the same, which is the fact established by the cross-modality matches. This leads to some interesting deductions about the way the Central Nervous System (CNS) operates.

Both the Fechner-Weber and the Stevens Laws are the result of empirical observations and have not hitherto been derived from any basic physical, physiological, or psychological considerations. In the next chapter we will look for reasons why these two laws take the form they do, at least in relation to olfaction.

Before leaving the subject of empirical psychophysics, it is useful to relate its two basic laws to several olfactory phenomena which have so far not been mentioned here.

One of the most curious of the many curious things about smell is that there are no characterless smells. Each is distinct and each retains its distinctiveness almost down to the threshold value. Right at the threshold there may be some indefiniteness which is revealed by remarks like, "... yes, I smell something but I'm not sure what... ". As soon as the strength is increased even very slightly (by one or two steps in Table 1, Chapter 16) the response changes to a positive, "... it's faint but I can recognize it... ". It is this persistence of olfactory quality right down to the lowest perceptible levels that enables properly trained dogs to follow a track or sniff out a hidden person or object.

Our eyes do not have this ability to recognize colors at low levels of illumination. The threshold for color recognition is many orders of magnitude above the threshold for just perceiving light in general. This is a result of our having two kinds of light-sensitive devices (rods and cones) in our retinas. The fact that our noses continue to discriminate even at very low concentrations is a consequence of the relatively primitive molecular mechanisms they employ. (There may, however, be something in our noses analogous to the rod and cone systems in our eyes, which will be considered in Chapter 21.)

While odors are distinctive even at the lowest intensities, they can nevertheless change their character progressively at higher concentrations. At high concentration the smell of a skunk is abominable, but when sufficiently dilute it is not unlike musk which is a common and sought-after component of many perfume formulations.

The change of odor quality with strength is probably due in most cases to the fact that the stimulus is a mixture of several different chemical substances. In the Stevens formula, the rate at which the smell sensation grows depends upon the constant, n, in the formula

$$S = kC^n$$

If the values of n are somewhat different for the different components of the mixture, one that tunes in late may yet grow relatively faster than another and come to dominate the pattern when the total concentration is high enough. In the same way, if the JND for one component is even slightly larger than that of another component of a complex stimulus, it will enable that compound to outstrip the other as the concentration increases.

Whether the same thing occurs with highly purified chemicals is not entirely clear because of experimental difficulties from adaptation and fatigue. In view of what was said in Chapter 16 about the influence of molecular symmetry on the process of interaction, it would not be surprising if the relative contributions of different vibrational modes could change with concentration because of a difference in the JND or Stevens exponent.

What is most significant is that the change of quality with concentration is never abrupt and it is usually possible to discern the essential quality over a large range of concentrations. A dog, for example, when allowed to sniff a person's cupped hands will then select an object which has merely been touched by one of that person's fingertips.

Evolutionary selection has operated to give this sensory modality a special role as a channel of communication between the organism and its environment.

REFERENCES

1. **Jones, F. N.,** Scales of subjective intensity for odors of diverse chemical nature, *Am. J. Psychol.,* 71, 305, 1958.
2. **Jones, F. N.,** Subjective scales of intensity for the three odors, *Am. J. Psychol.,* 71, 423, 1958.
3. **Stevens, S. S.,** Neural events and the psychophysical law, *Science,* 170, 1043, 1970.

Chapter 18

PHYSICS OR PSYCHOPHYSICS?

We are gifted with the power of imagination, and by this power we can enlighten the darkness which surrounds the world of the senses. Bounded and conditioned by co-operant reason, imagination becomes the mightiest instrument of the physical discoverer.

J. Tyndall

As explained in Chapter 12, the ability of an odorous molecule to interact with a receptor site appears to depend upon its ability to carry away a quantum of energy from the site and thereby to increase the ion-permeability of the membrane. The reliability of the system would be improved if each primary receptor neuron were to carry several sensitive sites and to fire only when several of them are triggered approximately simultaneously. Given this picture, both the Fechner-Weber and Stevens formulas as they apply to odorous stimulation can be deduced from elementary physics.

The olfactory epithelium consists of many millions of olfactory cilia, about ten of which emanate from the tip of each primary neuron. There may be as many as 30 or more *kinds* of neuron, each kind being selectively sensitive to a specific molecular-vibrational frequency in the stimulus. Let us consider the cilia emanating from neurons of one of these kinds.

The cilia from neurons of a given type present a total of n receptor sites, each with an area, a cm^2, so as to present a total area na cm^2 to the molecules of the stimulus. If there are C molecules of stimulus per cubic centimeter, the number of molecules impinging on na square centimeters per second, as calculated from elementary kinetic theory, is naCv/4, where v is the average speed of the molecules, which is 0.921 times the root-mean-square (rms) velocity. The rms velocity, u, is immediately calculable from the formula

$$u = \left(\frac{3RT}{M}\right)^{1/2}$$

where R is the gas constant, T is the absolute temperature, and M is the gram molecular weight. If the number of impacts required to discharge a neuron is g, and if a fraction, h, of the impacts is effective for steric or other reasons, the number of sites which become desensitized in 1 sec will be

$$\frac{naCvh}{4g}$$

If the fraction of n which is in the desensitized state at any instant is b, a fraction $(1 - b)$ is receptive to new impacts, and so the rate at which new cells are stimulated is

$$\frac{n(1 - b)\,aCvh}{4g}$$

The rate of recovery will be proportional to the fraction, b, which has become desensitized, or

$$tbn$$

where t is a proportionality constant having the dimensions of time^{-1} and may be regarded as the reciprocal of the time required for a receptor site to recover after interacting with a stimulus. This is usually about 10 msec so that t may provisionally be taken as equalling about 100 sec^{-1}.

A steady state is reached when the rate of desensitization equals the rate of recovery, that is, when

$$\frac{n(1 - b)\,aCvh}{4g} = tbn$$

from which

$$b = \frac{aCvh/4g}{aCvh/4g + t}$$

Physically, b represents the fraction of the neurons which are firing at any given instant, that is, it is the signal moving into the Central Nervous System (CNS) from receptors of the particular type under consideration.

In any given experiment, a, v, h, and g, are all constant. If aCvh/4g is very much larger than t, then

$$b = \text{constant}$$

and the signal is independent of the stimulus concentration. This happens when there are enough stimulus molecules to saturate the receptors. At the other extreme, if aCvh/4g is very much smaller than t, then

$$b = \frac{aCvh}{4gt}$$

and the signal is directly proportional to C. For intermediate conditions, S is proportional to C^x, where x lies between 0 and 1. A recent critical review of the data for 110 substances gives values of x ranging from a low of 0.12 to a maximum of 0.87.

Equation 1 specifies only one part of the input arriving in the CNS. The olfactory nerves constitute what information theorists call a noisy channel because nerves do occasionally fire spontaneously and the CNS must somehow distinguish between genuine signals and random noise. The best way to maintain a constant signal-to-noise ratio when C changes is for the sensation, S, to be related to b, the rate of arrival of genuine signals, by the formula

$$\text{Log}S = b$$

If the sensation, S, is measured by the method of ratio estimation, we have, empirically

$$S/S_0 = (C/C_0)^X$$

from which

$$\text{Log}(S/S_0) = x\text{Log}(C/C_0) = b_c - b_0$$

or, in general

$$x = \frac{b_2 - b_1}{\text{Log}(C_2/C_1)}$$

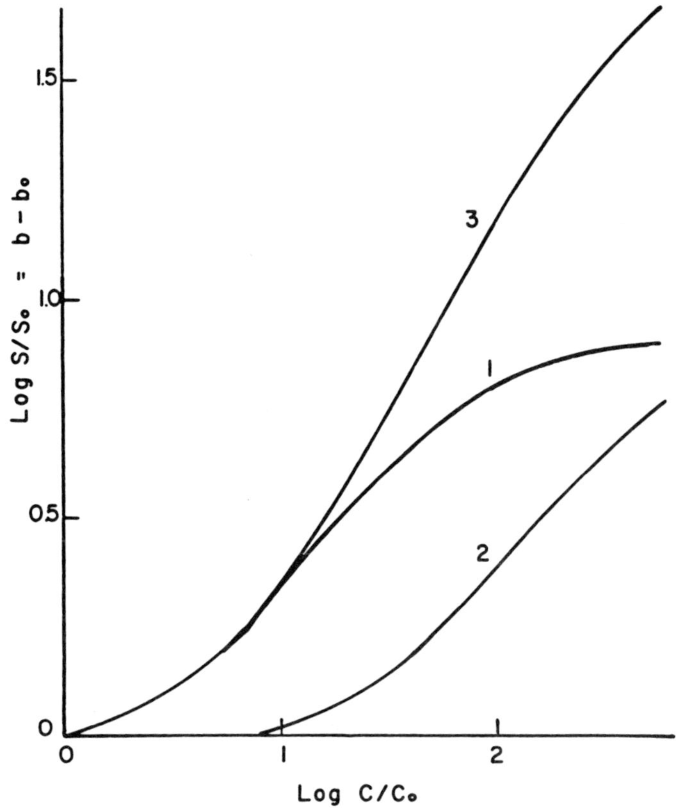

FIGURE 1. Curve 1 is the Log-Log or Stevens Plot of odor strength from a single type of receptor if S is measured by the method of magnitude estimation with various concentrations of stimulus, C. Curve 2 is the input from a second type of receptor with a higher threshold, and Curve 3 is the combined odor strength from the two inputs.

The case of benzene, mentioned in the preceding chapter, where $x = 0.52$, may be taken as representative. For benzene, the threshold concentration is reported to be 3.65×10^{13} molecules per cubic centimeter. With a molecular weight of 78, the average molecular velocity at 300 K is 2.92×10^{4} cm/sec. We can only guess at the values of the terms a, h, and g in Equation 1, but on the basis of what has been said earlier it is reasonable to suppose that the area, a, will be quite small, perhaps only a few square Angstroms or, say, 4×10^{-16} cm². We can also put $g = 4$ and $h = 0.4$ and see what happens. Doing so, we have, for benzene

$$aC_0 vh/4g = 1.14$$

If $t = 100 \text{ sec}^{-1}$, from Equation 1,

$$b_{100} = \frac{114}{214} = 0.532$$

$$b_{10} = \frac{11.4}{111.4} = 0.102$$

and, from Equation 2,

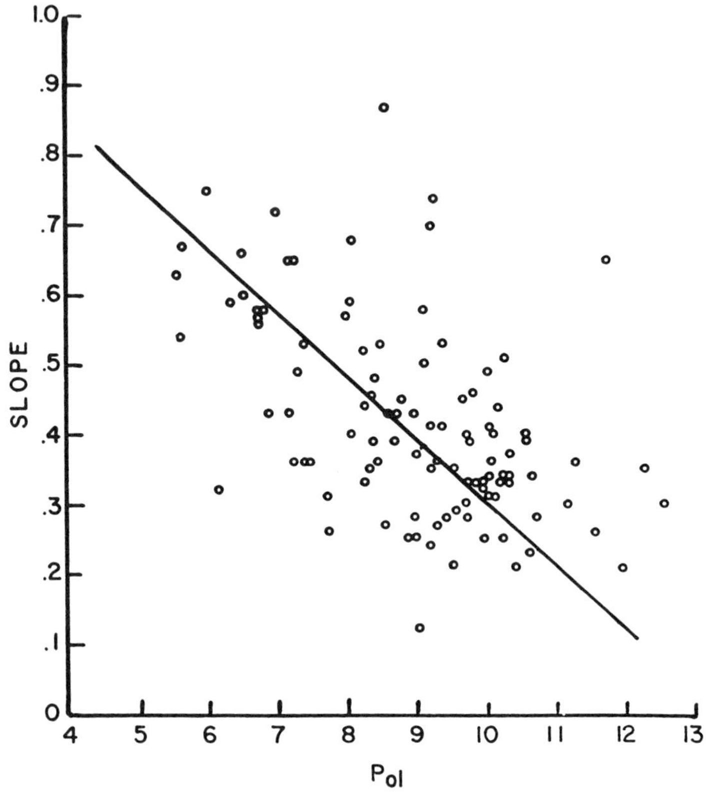

FIGURE 2. The plot of "slope" (Stevens exponent) against −Log (Threshold) or P_{ol}, indicates some degree of correlation between the two. For an explanation of the large amount of scatter, see Figure 3.

$$\frac{b_{100} - b_{10}}{\text{Log}(100/10)} = \frac{0.532 - 0.102}{1} = 0.43$$

This is the predicted value of the exponent, x. In the last Chapter, Jones' experimental value was 0.52. A more recent critical compilation suggests a still higher value of 0.65. It must be remembered, however, that this calculation gave the estimated input from a single type of receptor. The actual sensation results from the stimulation of several selectively sensitive types of receptor which need not all have the same threshold sensitivities.

In Figure 1, Curve 1 shows the calculated input from a single receptor type, assuming the numerical values used above. Curve 2 shows the input from a second receptor with a somewhat higher threshold, and Curve 3 shows the combined inputs. Evidently the straight part of the plot is much longer with the combined inputs. The slope of the curve, represented in psychophysics by the exponent, x, will be somewhat dependent on how soon the various inputs make their contribution.

The efficiency of the interaction between stimulus and receptor is represented in Equation 1 by the threshold concentration, C_0, which we showed in Chapter 12 is related to the turnaround time between successive interactions by the stimulus. The additional terms, h and g, and especially their ratio, h/g, constitute a further factor in determining the interaction efficiency. In the above example, h/g was assigned a value of 0.1. Giving it a value of 0.2 would change the exponent, x, to 0.51 and reducing it to 0.05 would make x = 0.31.

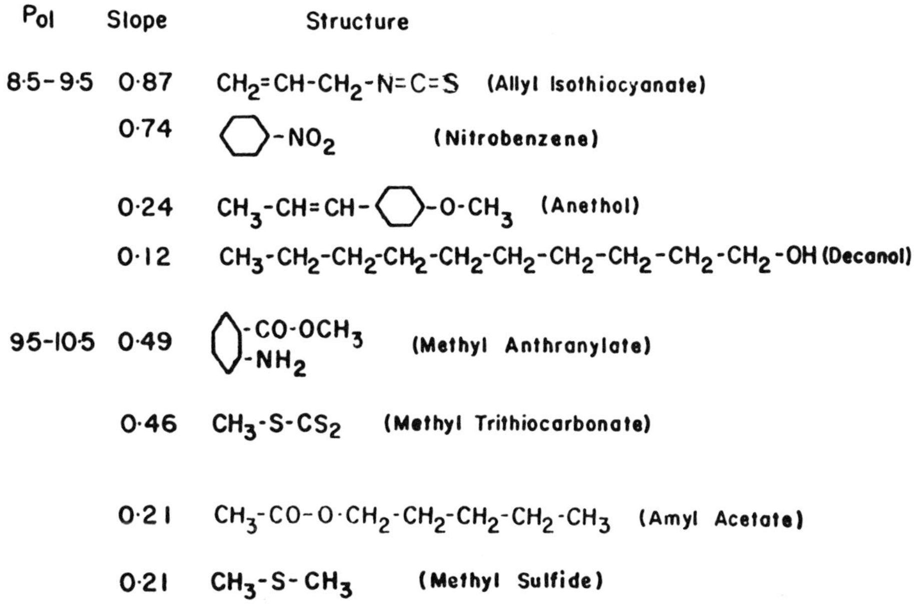

FIGURE 3. Compounds with approximately the same P_{ol} and with, respectively, the highest and lowest slopes are compared. High slopes seem to be associated with relatively inflexible structures and low ones with molecules that would probably flex rather easily when in the near vicinity of a receptor site.

This is an interesting result. What it suggests is that a high value for the Stevens exponent indicates an osmically efficient molecular stimulus, and vice versa. The point merits a closer look.

Mention has been made of the critical compilation of thresholds and slopes (that is, the Stevens exponents) for 110 compounds. Figure 2 shows a plot of the slope against P_{ol}, P_{ol} being the negative logarithm of the threshold concentration expressed in gram moles per liter of air. Evidently there is a tendency for low values of the exponent to be associated with high values of P_{ol} but there is a great amount of scatter on either side of the mean. If we look at the highest and lowest values of the slope for a given range of P_{ol}s, Figure 3 shows that high values of the slope are associated with molecular structures that seem to be relatively inflexible, and that low values seem to correlate with rather flexible structures. There does not seem to be any direct way to measure molecular flexibility but it can be inferred indirectly from the heat capacity per gram mol.

In a gas at constant pressure the molecular heat capacity is made up of contributions from three degrees of freedom of translational movement, two of rotation, a variable number of active intramolecular vibrations, and a term for the work done in expanding against a constant external pressure. In a liquid the situation is qualitatively similar, but quantitatively more complex because of the close packing of the molecules. One consequence of this is a substantial contribution to the observed heat capacity from the work done in expanding against the strong cohesive forces of the liquid. Nevertheless, it seems clear that flexible molecules will owe their flexibility to relatively weak restoring forces and so will have relatively many vibrational modes with frequencies low enough to be active at the ordinary temperature. By contrast, a stiffer molecule with stronger force constants in its elastic structure, will absorb less heat into its vibrational modes when the temperature is raised one degree.

Hence, other things being more or less equal, the molecular heat capacity of a flexi-

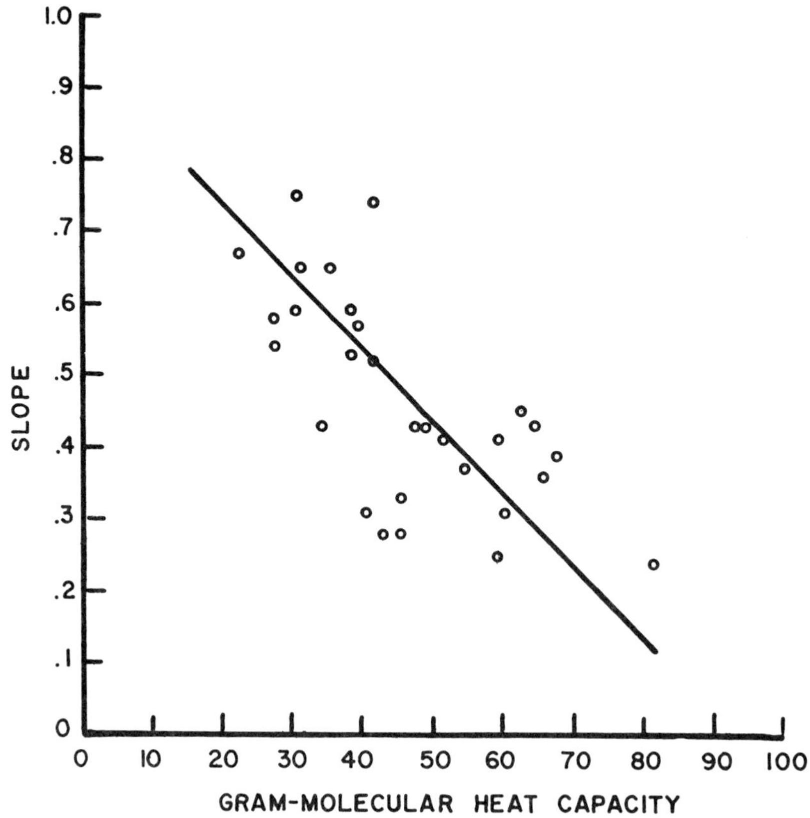

FIGURE 4. The gram molecular heat capacity of a liquid is a rough measure of the molecular flexibility: the greater the flexibility the larger the heat capacity. These results verify the expected correlation between molecular stiffness and stimulatory efficiency as represented by the Stevens exponent.

ble molecule should tend to be greater than that of a more rigid one, even though measured in the liquid state. And if so, the fact that flexible molecules are less efficient stimulators and have lower slopes would suggest that such molecules would also have higher molecular heat capacities. Figure 4 shows Stevens slopes and molecular heat capacities for those compounds for which both quantities are readily available. The expected correlation is at once apparent.

In Chapter 15 we saw that the stimulus molecules may be in some measure twisted out of shape in the near vicinity of a receptor. Rigid molecules would preserve their specific vibrational character during a close approach much more than would flexible ones.

The foregoing analysis related to the Stevens form of the psychophysical law, and evidently the theory developed in the earlier Chapters affords some interesting insights into the physics of psychophysical magnitude-estimates.

The situation is equally clear with the Fechner-Weber form of the law.

As before, we start with the expression

$$\mathrm{Log}S = b$$

which can also be written

$$S = \mathrm{Log}^{-1} b$$

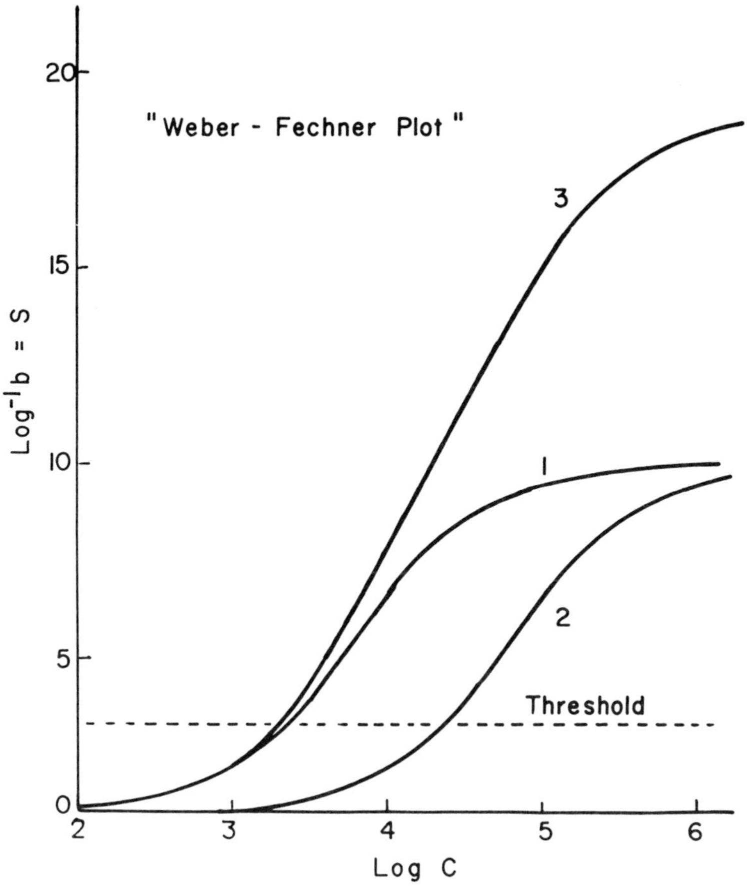

FIGURE 5. The Fechner-Weber or semi-Log Plot of odor strength, if S is measured in terms of Just Noticeable Differences or after the manner of Figure 1, Chapter 16. Curves 1 and 2 are for single types of receptor with different thresholds, and Curve 3 is the sensation from the combined inputs.

or, in English: S is the number whose Logarithm is b. Note, however, that S is now measured by the kind of experiment that produced Figure 1, Chapter 16. There it was shown that

$$S = k\text{Log}C$$

but if

$$S = \text{Log}^{-1} b$$

a plot of Log^{-1}b against LogC should give a straight line.

Figure 5 shows S plotted against LogC for a single type of receptor in Curve 1; for a second type with a higher threshold, Curve 2; and the combined input from the two receptors in Curve 3. The curves are straight in the middle range but flatten out at their extremities as do many experimentally determined ones.

What is especially interesting is the way Figure 5 shows how it is that subthreshold stimuli can make a contribution to the perceived sensation.

REFERENCES

1. **Guadagni, D. G., Buttery, R. G., Okano, S., and Burr, H. K.,** Additive effect of sub-threshold concentrations of some organic compounds associated with food aromas, *Nature (London),* 200, 1288, 1963.
2. **Langmuir, I.,** The constitution and fundamental properties of solids and liquids, *J. Am. Chem. Soc.,* 38, 2221, 1916.
3 **Neuhaus, W.,** The ability of dogs to discriminate mixtures of odors, *Z. Vgl. Physiol.,* 39, 25, 1956.
4. **Patte, F., Etcheto, M., and Laffort, P.,** Selected and standardized values of suprathreshold odor intensities for 110 substances, *Chem. Senses Flavor,* 1, 283, 1975.
5. **Wright, R. H.,** The perception of odor intensity: physics or psychophysics?, *Chem. Senses Flavor,* 3, 73, 1978.
6. **Wright, R. H.,** The perception of odor intensity: physics or psychophysics. II, *Chem. Senses Flavor,* 3, 241, 1978.

Chapter 19

NEURAL CODING — I

Hypothesis is the general, and experiments are the soldiers.

Leonardo da Vinci

In Chapter 10 we moved from the outside world of stimulus molecules up to the threshold where these molecules make contact with the substance of the organism itself. Let us now move inside the organism to look at the information the molecules bring and while it is in transit from the primary receptors to the Central Nervous System (CNS).

The picture projected in Chapter 2 and developed in the succeeding chapters can be summarized in Figure 1. A molecule with a specific set of attributes represented by the letters H, L, and M, skips across an array of receptors some of which are sensitive to the H, L, and M attributes while others are sensitive to other attributes not carried by this stimulus and not, therefore, stimulated by it. In the picture, all the receptors of type H are connected by nerve fibers to a structure in the brain which lights up in some way when the visiting molecule has the attribute, H. This happens also with the L and M attributes, but not with the others. In this way, a pattern of nerve impulses is presented to the brain which corresponds in a one-to-one fashion with the pattern of vibrational attributes presented by the molecules of the stimulus.

This is clearly a highly simplified picture, if only because it takes no account of the possibility that the intensities of the incoming Hs, Ls, and Ms may be different. Nevertheless, it is a picture that is clear enough to be compared with the results of experiment.

By attaching microelectrodes to single nerve fibers, electrophysiologists have been able to pick off the information passing along them and display it on a cathode-ray oscilloscope. When this is done, it turns out that every time a neuron fires, an action potential lasting about a millisecond and looking like a spike on the oscilloscope trace passes along the nerve with a velocity rather less than the velocity of sound in air. The size of the spike, that is, the amplitude of the voltage swing, is the same for a strong stimulus as for a weak one. A strong stimulus merely sends along more action potentials, up to a maximum set by the duration of the refractory period which follows the stimulation of each spike. It is this simple alternative, fire-or-don't-fire that makes the kind of yes-or-not-yes signalling system described in Chapter 2 so directly applicable to the olfactory transmission of information.

There is evidence that each action potential in an olfactory neuron is the result of one interaction between the cell and one passing stimulus molecule.

Several experimenters have inserted microelectrodes into the olfactory epithelia of frogs or rats and recorded the action potentials passing along single nerve axons when various odorous stimuli were presented to the organism. Much of the work was done before the vibrational theory had been developed to its present level, and so the results of the electrophysiological work were often difficult to interpret.

The principal difficulty (there are others to be discussed later) was the confusing way each single nerve responded to different chemical stimuli. The following is a simplified example of the sort of situation that emerged.

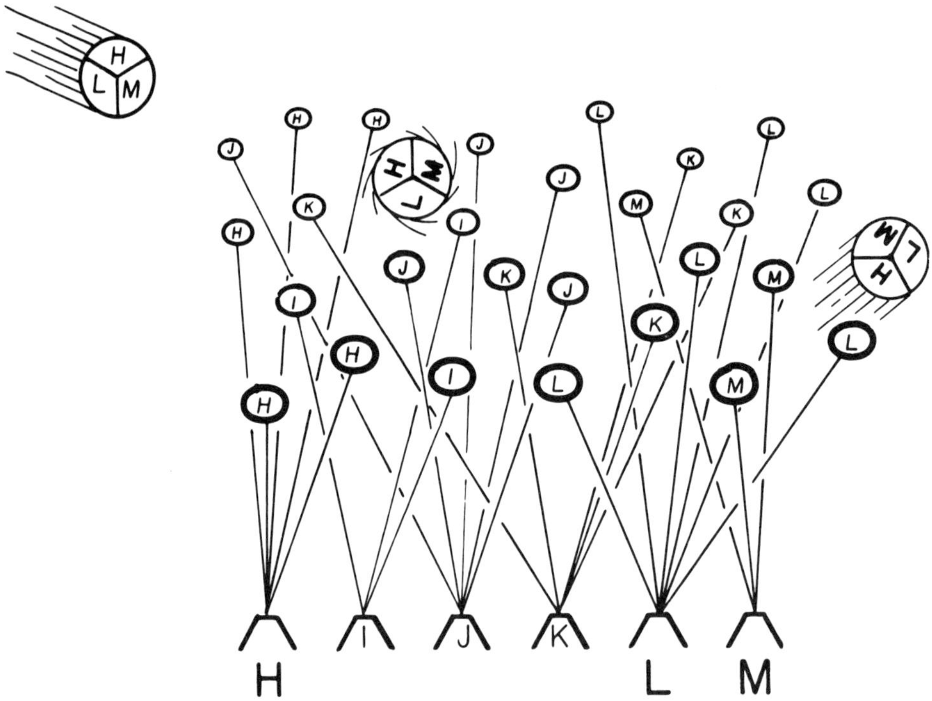

FIGURE 1. A molecule with a combination of attributes represented by the letters H, L, and M, moves over the olfactory receptor surface and interacts with selectively sensitive receptors so as to inject an H-L-M signal into the Central Nervous system.

Nerve number:		1	2	3	4	5	6
Stimulus	A:	+	+	−	−	+	−
	B:	+	−	+	−	+	+
	C:	−	−	−	+	+	+
	D:	−	−	−	−	+	+

The first nerve to be tapped responded to stimuli A and B only. The next one responded to A but not B. Number 3 responded to B but not A, and number 4 responded to C only while number 5 responded to all four stimuli, and number 6 to B, C, and D, but not A.

This sort of thing led Lettvin and Gesteland, who pioneered in this extremely tricky experimental technique, to view the results as presenting a picture of complete anarchy in which "every fibre has a different point of view from every other fiber". Later workers have tended to agree generally with this, though not, perhaps, in such extreme terms. Before the difficulty could be sorted out it had to be realized that the molecules of each stimulus present *patterns* of vibrational attributes, and that different molecules (with different odors) may have certain frequencies in common. Molecules with a shared frequency will therefore stimulate nerve cells tuned to that frequency. A nerve that is tuned to some other frequency which one kind of molecule has and the other lacks will respond to the one but not to the other, and so on.

This essentially simple way to interpret the results has been delayed by two of the minor difficulties alluded to above.

The first was the fact that nerve number 5 (in the hypothetical example above) might respond strongly to Stimulus B and much less strongly to C, where number 6 might respond equally to both and in still another the relative intensities of the responses might be reversed. If a difference in the strength of the response (that is, in the number of action potentials per second) is taken as indicating a difference in the specificity of the receptor, then the picture is indeed one that approaches complete anarchy. The fact is, however, that vibrational modes can have nearly the same frequency in different molecules but if the physical nature of the movements differ (have different symmetries) then the stimulatory effectiveness will be different even though the frequencies are the same and the molecules are presented at the same concentration. This was the situation allowed for by introducing the term, h, in the formula developed in the last chapter. What matters is that the cell responds significantly. How strongly it responds is secondary.

The second difficulty appeared when the techniques of single-fiber recording were progressively improved. It then appeared that there is a certain amount of spontaneous firing of the neurons even in the apparent absence of any stimulus. When a stimulus is presented, these basic, spontaneous discharges may be, and usually are, increased or excited, but sometimes they are decreased or inhibited. This was taken as showing that the olfactory code is not a binary, "yes-or-not-yes" one, but rather a ternary, "yes-or-no-signal-or-no" one. There are several reasons why this may be an unnecessary complication in interpreting the phenomena. If a single action potential is generated by a physical quantum-interaction, the stimulating molecule will usually leave the receptor site during the few milliseconds of the refractory period which follows the interaction. At times, and for some stimulus molecules, there may be some chemisorption which delays the escape of the stimulus and prolongs the refractory period. Thus, although the inhibitory responses are a real feature of the electrical recordings they may not, in fact, contribute in an important way to the olfactory information.

All neurons fire spontaneously at irregular intervals and the receptor cells are no exception. The organism must therefore disregard the spontaneous signals and respond only to the augmentation produced by an external stimulus or, of course, the diminution following the removal of the stimulus. If the organism ignores the spurious signals, from an informational standpoint receptors whose spontaneous firing is inhibited are the same as those whose spontaneous firing is consistently ignored.

In the next chapter we will see how the purely excitatory responses of different neurons can be provisionally correlated with the vibrational attributes of the molecules used as stimuli.

Before doing so, however, it is necessary to clear up a possible source of misunderstanding. Figure 1 appears to depict the stimulus molecules as bounding over an essentially dry array of receptor sites, whereas the encounters are actually with receptor sites bathed in a fluid medium from which positive ions move into the cell interior when the membrane is opened. The contradiction is more apparent than real.

The vertebrate olfactory epithelium terminates outwardly in a tangle of slender cilia which may be as much as 0.1 to 0.2 mm long. For part of their length they may float half in and half out of the liquid mucus, held there by surface forces. The phenomena are well-known in a wholly unrelated field of technology, namely, mineral dressing.

In the process of concentrating minerals by froth flotation, finely powdered mixtures of ore particles and waste gangue are suspended in water containing a special oily additive. When air is blown through the mixture, the oil coats the bubbles and preferentially wets the particles of ore but not the gangue so that the ore particles lie half in and half out of the oily film on the bubbles and are selectively floated out of the mixture. An exactly similar mechanism could cause the distal parts of the cilia to float in the surface, but with a wetted film no more than a few molecules thick bathing the part in contact with the air.

A molecule with a diffusivity of 10^{-5} cm^2 sec^{-1} would diffuse through a film 0.001 mm thick in about 5×10^{-4} sec, so that the two pictures are not mutually contradictory.

Even though an insect's antennae appear to be dry, in contrast with the film of mucus in the terrestrial vertebrate epithelia, there may in fact be a film of fluid bathing the outside of the receptor sites. Since most insects needs must resist desiccation, some means must exist to limit the evaporation of water from such films. The necessary mechanism is well established: vapor pressure over a concave surface is less than from a flat surface, the effect being greater the smaller the radius of curvature. By placing the sensitive sites at the bottoms of small openings or pits in the cuticle, the liquid surfaces will be curved in just the way needed.

It may be pointed out also that the interaction mechanism as outlined will work as well with fish as with air-breathing insects, reptiles, or mammals — once again in conformity with its primary position in the evolution of sensory modalities. Where fish are concerned, the need for volatility is replaced by the need for water solubility. Hence it is possible for nonvolatile, and to our noses odorless, amino acids to act as olfactory stimuli for fish in a way to be described in Chapter 41.

Chapter 20

NEURAL CODING — II

The greater the number of facts consistently explained by one and the same theory, the greater the probability of its being true.

J. W. Mellor

When electrophysiological recordings are made from a single receptor neuron in the olfactory epithelium of a vertebrate or the antenna of an insect, how can the frequency to which that neuron is tuned be discovered? The question is a vital one. If it can be answered we will be able to correlate directly the physical attributes of the outside stimulus with the information passing into the Central Nervous System (CNS) via the olfactory nerves.

To answer the question we require data for the response of a single receptor unit to each of several chemicals with known vibrational characteristics. Unfortunately, many electrophysiologists have used rather concentrated stimuli so as to ensure that the units will respond. Under these conditions there will be a high level of overall activity, possibly accompanied by some interfiber stimulation, especially where there has been damage from the insertion of the recording microelectrodes. Therefore, useful correlations between molecular attribute-patterns and neural discharge-patterns will be clearly discernable only when the stimuli are presented at concentrations comparable to those that elicit normal behavioral responses. It is necessary also that the data be reported in enough detail for the single-unit response in the neuron to each chemical to be identified. And, of course, the vibrational characteristics of each stimulus must be known.

At the time of writing, not all these conditions have been adequately met so that one has to be content with results that are more suggestive than convincing.

Thus, D. L. Blank[1] reported single-unit recordings from primary olfactory neurons of the grass frog, *Rana pipiens,* and his battery of test stimuli included formic acid, FA, cineol, CIN, menthone, MEN, and 15-hydroxypentadecanoic acid lactone, PDAL, for which absorption spectra were available in my file in the range 500 to 100 wave numbers. This embraces most but probably not all of the vibrational frequencies likely to be significant.

In the upper part of Figure 1 the positions of the absorption maxima of these substances are plotted against the grid of computed receptor frequencies based on the formula, $F = 6.4N'$ derived in Chapter 9. A compound with an absorption maximum near one of these computed frequencies should be at least potentially capable of stimulating a receptor with that frequency-response. Where an absorption peak falls approximately midway between two grid lines it may interact with either or both of the adjoining receptors, bearing in mind that the actual positions of the absorption maxima are subject to some uncertainty.

In the lower part of the figure, the recordings from the various nerve fibers (which were identified by number) are grouped on the basis of their excitatory responses, regarding no-response (0) and inhibition (−) as equivalent from the informational point of view, given an essentially binary code. Thus, for example, the Units numbered 7 and 14 in Blank's report are similar in showing positive responses to only one formic acid (FA) of the four stimuli under scrutiny here. Numbers 3 and 8 responded to CIN and PDAL but not to the other two, and so on. The fact that Unit 7 responded to FA with intensity 6, and Unit 14 with intensity 2, is not as important as the fact that they both showed this degree of selectivity.

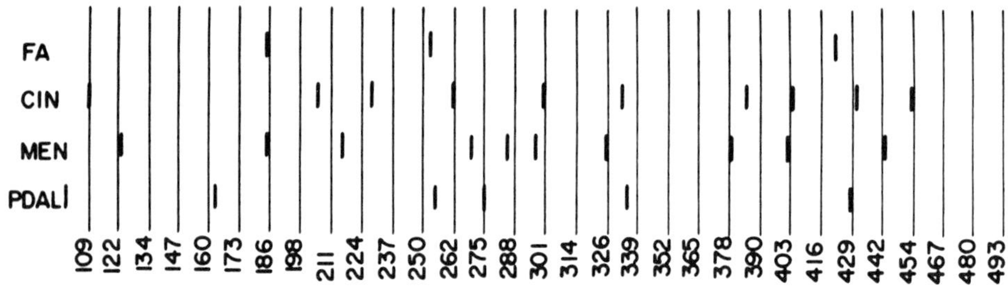

Unit:-	1	7, 14	6	5,10,13	9	12,19,20	3,8	4	16	2	17	18,22	11	15,21
FA	0	6 2	0	- - -	1	1 1 4	0 0	0	0	1	0	5 3	2	3 3
CIN	0	0 -	0	0 0 -	0	0 - -	2 1	0	4	1	6	6 2	0	6 1
MEN	0	0 0	2	- 0 0	5	- - -	0 -	4	3	0	1	1 2	2	2 2
PDAL	0	0 -	-	4 2 2	-	1 3 6	2 3	1	·	1	1	6 3	6	0 0
Band:-	147 173 198 etc	416	122 288 378 442	160	186	250	339	275	211 301 403	429	326	262	?	?

FIGURE 1. The positions of the far infrared absorption maximum of formic acid, FA, cineol, CIN, menthone, MEN, and 15-hydroxypentadecanoic acid lactone, PDAL, are shown in relation to the grid of frequencies to which receptor cells may be tuned. In the lower part, the arbitrarily numbered receptor units in the frog's olfactory epithelium are grouped on the basis of their positive (excitatory) responses to these stimuli and given provisional frequency-assignments based on nearness to absorption maxima in the stimuli that excite them. (From Wright, R. H., *J. Theor. Biol.*, 64, 473, 1977. With permission.)

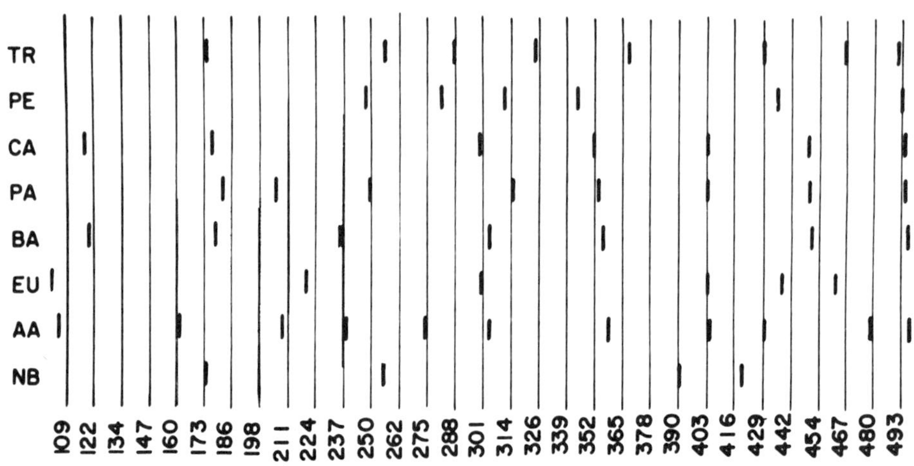

FIGURE 2. The positions of the far infrared absorption maxima of eight stimuli used to excite single receptor units in the male silkmoth, plotted against the grid of receptor frequencies. TR = terpineol; PE = phenylethyl alcohol (beta); CA = cinnamaldehyde; PA = phenethyl acetate; BA = benzyl acetate; EU = eugenol; AA = amyl acetate; NB = nitrobenzene. (From Wright, R. H., *J. Theor. Biol.*, 64, 473, 1977. With permission.)

Taken together, the upper and lower parts of Figure 1 suggest that Units 7 and 14 are sensitive to a frequency near 416 cm^{-1}. Unit 6, which responded to MEN only, could be sensitive to any one of four possible frequencies: 122, 288, 378, or 442 cm^{-1}.

Table 1
FREQUENCY ASSIGNMENTS FOR 28 OF THE 48 STIMULI USED WITH *A. PERNYI*, AS DEDUCED FROM FIGURE 2

Unit:-	17,19,26,29,30,35,36,37,41,44	18	8,9,10,38,39	23	31,32	22	40	16,20,33,46	43	7,24
TR	0 - - nd nd - - - 0 0	0	1 3 1 1 3	2	3 2	0	0	2 2 2 2	2	2 1
PE	0 - 0 0 0 - - 0 0 0	0	0 0 0 - -	-	- -	-	0	1 1 2 2	0	1 1
CA	0 - 0 0 0 - - 0 0 0	0	0 - - - -	0	- -	-	0	0 0 nd 0	-	1 1
PA	- - - - 0 - - 0 0 -	0	- - - - -	-	- -	-	1	0 - 0 0	-	3 3
BA	- - - - 0 - - 0 0 -	-	- nd nd - -	-	- -	-	0	0 - - 0	1	2 3
EU	0 - - - - - - - 0 -	0	0 - - - -	0	- -	2	0	0 - - 0	-	0 -
AA	- - 0 0 nd 0 0 0 0 nd	1	nd - - 0 0	0	2 1	-	1	0 - nd nd	0	nd 1
NB	0 nd - - - - - - 0 -	0	nd - - - -	2	- -	-	0	0 nd - nd	-	nd -
Band:-	135, 147, 390, or 416	160 or 480	288 or 326	262	365 or 429	224	211	288	467	493

Note: O = no responds, — = inhibitory response, integers indicate the approximate strength of a positive (excitatory) response. nd = not determined, i.e., that stimulus not applied to that unit.

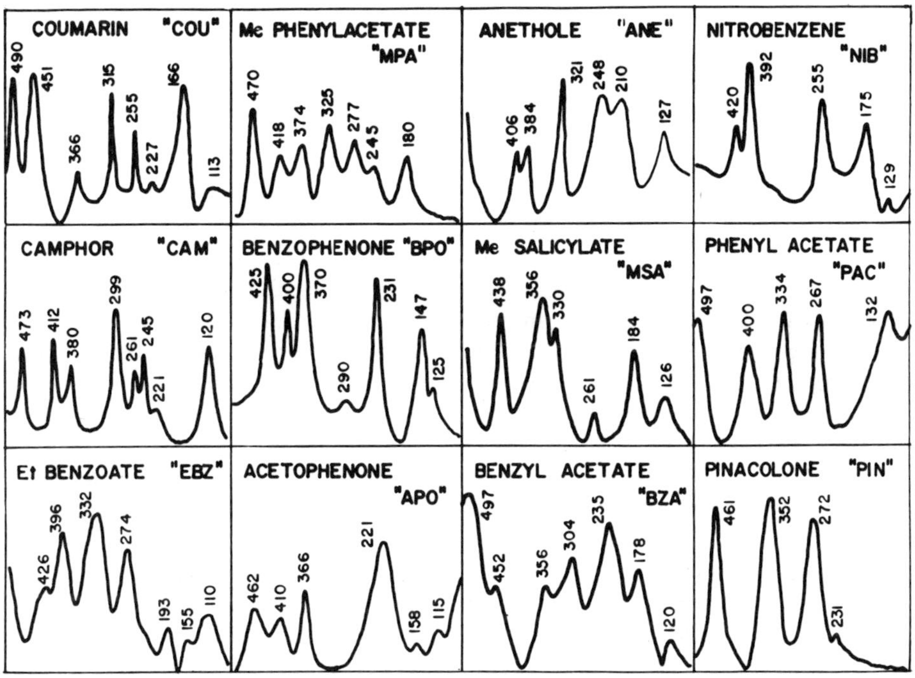

FIGURE 3. The far infrared absorption spectra of substances whose availability and well-defined absorption maxima should make them especially suitable stimuli for use in deducing receptor frequencies from electrophysiological recordings. (From Wright, R. H., *J. Theor. Biol.*, 64, 473, 1977. With permission.)

More data with more stimuli of known vibrational character would be needed to narrow down the choices.

These results are clearly no more than a first approach to the problem, but they

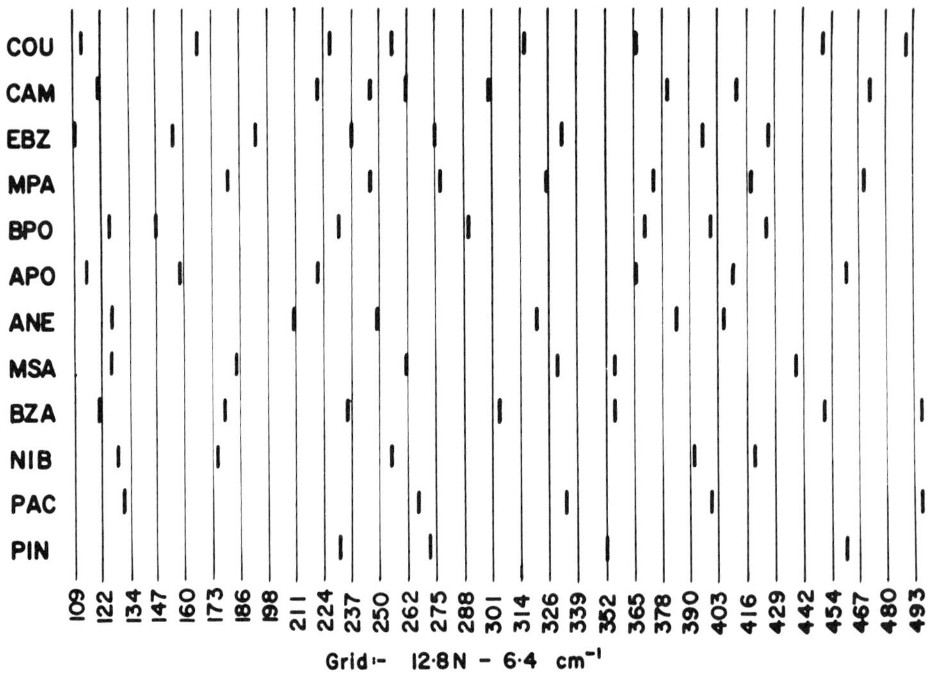

FIGURE 4. The vibrational frequencies of the recommended stimuli shown in relation to the grid of receptor sensitivities. A receptor tuned to 352 cm^{-1} would be expected to respond to MSA, BZA, and PIN but would be unlikely to respond to the other stimuli shown, and one tuned to 288 cm^{-1} should respond selectively to BPO, and so on. (From Wright, R. H., *J. Theor. Biol.*, 64, 473, 1977. With permission.)

show that frequency-assignments can be made in a provisional way to all but 3 of the 22 neurons from which recordings were reported. Since the range from 500 to 100 cm^{-1} does not necessarily include the whole range of osmically active frequencies, it is hardly to be expected that every receptor unit will be assigned a frequency, and some of the assignments would doubtless be altered given more complete data. The essential purpose of Figure 1 is to show that the responses of the various primary receptors are capable of being plausibly correlated with the vibrational attributes of the stimuli which elicit them.

A more detailed test can be made using data reported by D. Schneider and his colleagues[2] based on recordings from the antennal receptors of the male silkmoth, *Antherege pernyi.*

Here, again, not all the receptor units were exposed to all the chemical stimuli, nor are far infrared spectra available for all the chemicals they used. Figure 2 shows the spectral absorption maxima for eight of their stimuli plotted against the calculated grid of frequencies. Table 1 shows that provisional frequency assignments can be made for 28 of the 48 units from which recordings were reported. Of the remaining 20, 11 were exposed to so few of the stimuli that they could not be included in my analysis, and the 9 remaining could not be given assignments in the range 500 to 100 cm^{-1}.

This evidence is too partial and incomplete to be convincing, but it is suggestive of an approach more likely to be fruitful than the conclusions expressed heretofore: "the raw coding data . . . baffles the decipherer",[8] or, "the single unit electrophysiological picture that has been emerging . . . suggests that the simplistic codes are not to be found,"[1] or, ". . . every given olfactory fiber displays its own spectrum of sensitivity".[7]

In future experiments of this kind it would be desirable to employ stimuli with known and well-defined vibrational characteristics. Figure 3 shows the spectra of 12 compounds chosen on the basis of stability, availability, and sharply-defined far infrared absorption maxima, which are the properties needed to make them suitable for this sort of work. Figure 4 shows their peak frequencies plotted against the presumed grid of receptor sensitivities. A nerve which responds only to benzophenone and none of the others could be given a frequency assignment of either 147 or 288, and the choice between the two made after a second round of tests using stimuli presenting one but not the other of these frequencies. In the same way, a unit which responds to camphor and benzyl acetate and to none of the others would probably be responding to a frequency near 301 cm^{-1}, and so on.

It is to be hoped that someone will be moved to make a systematic compilation of the far infrared spectra of the purest possible odorous substances, and preferably with the substances in the vapor state. The file in the possession of the writer (who is now retired and without access to the necessary laboratory facilities) was accumulated at various times and for various reasons, and not all the spectra are of the quality needed for this kind of work. Nevertheless, they have served to bring matters to the present position on the threshold of a new and clearer understanding of the phenomena.

REFERENCES

1. **Blank, D. L.,** Mechanism underlying the analysis of odorant quality at the level of the olfactory mucosa. II. Receptor selective sensitivity, *Ann. N.Y. Acad. Sci.,* 237, 91, 1974.
2. **Boeckh, J., Kaissling, K. E., and Schneider, D.,** Insect olfactory receptors, *Cold Spring Harbor Symp. Quant. Biol.,* 30, 263, 1965.
3. **Boeckh, J.,** Reactions of olfactory neurons in the insect déutocerebrum as compared to response patterns of olfactory receptor cells, *J. Comp. Physiol.,* 90, 183, 1974.
4. **Döving, K. B.,** An electrophysiological study of odor similarities of homologous substances, *J. Physiol.,* 186, 97, 1966.
5. **Döving, K. B., Nordeng, H., and Oakley, B.,** Single unit discrimination of fish odors released by char (*Salmo alpinus* L.) populations, *Comp. Biochem. Physiol.,* 47A, 1051, 1974.
6. **Gesteland, R. C., Lettvin, J. Y., Pitts, W. H., and Rojas, A.,** Odor specificities of the frog's olfactory receptors, in *Olfaction and Taste. 1,* MacMillan, New York, 1963.
7. **Holley, H., Duchamp, A., Revial, M. F., and Juge, A.,** Qualitative and quantitative discrimination in the frog olfactory receptors: analysis from electrophysiological data, *Ann. N.Y. Acad. Sci.,* 237, 102, 1974.
8. **Lettvin, J. Y. and Gesteland, R. C.,** Speculations on smell, *Cold Spring Harbor Symp. Quant. Biol.,* 30, 217, 1965.
9. **Wright, R. H.,** Odor and molecular vibration: neural coding of olfactory information, *J. Theor. Biol.,* 64, 473, 1977.

Chapter 21

ANOTHER KIND OF RECEPTOR

We must fight against the soporific influence of sham definitions. Especially must we guard against allowing them to stand in the way of an enquiry into facts.

A. Sidgwick

Up to this point my attention has been directed toward receptors which will respond selectively to any molecule with the correct profile which presents a vibrational mode close to the specific frequency-sensitivity of the cell. Since many molecules, but not all molecules, can fulfill this requirement, such cells are appropriately termed "odor generalists", and the ability of the organism to discriminate between odors depends upon which subset of its battery of odor generalists is stimulated.

In several species of insect, odor specialist cells have been described which show a much more specific and discriminating response, most often to a specific chemical secreted as a sex pheromone by a member of its own species and of the opposite sex. The existence of the two types of receptor can be accounted for as follows.

The dendrites of a generalist unit would carry several sensitive sites all tuned to the same frequency and therefore potentially responsive to any molecule carrying that frequency, provided enough sites interact within a sufficiently short time interval. A specialist cell could work in the same way except that it would carry sites of more than one frequency-specificity. A molecule capable of interacting with only one kind of site could not elicit a discharge, but one with the appropriate set of frequencies could.

The Lambert-Salter Plot in Figure 1 Chapter 10 shows that for molecules with a fairly low ν_{min}, say one below 200 cm^{-1}, fewer than 100 collisions are needed for a V-T relaxation. With a mean collision frequency of about 10^9 collisions per second, a stimulus molecule would need only about 10^{-7} sec. between site interactions. The diffusivity of a typical odorous molecule in air is about 0.1 cm^2 sec^{-1}, making it a very slow process over any but the shortest distances, and it is slower by a factor of about 10^4 in an aqueous medium. Nevertheless, even in water the molecule could traverse a root-mean-square (ms) distance of 1000 Å in about 10^{-8} sec, and in air the same distance in a time a thousand times smaller. We know nothing about the average distance separating the sensitive sites on the receptor cilia, but it seems clear that kinetically there is no reason why a molecule which finds itself in the vicinity of one site should not have a good probability of making the most of its opportunities and interacting repeatedly in times of the order of milliseconds, as described in Chapter 18.

This would be a by no means improbable adaptive development in the evolution of the receptor system.

In the process of organic evolution, this type of effect would become biologically significant as soon as multicellular organisms appeared with enough cell-specialization to make possible a rudimentary nervous system and long before anything resembling sight or hearing could evolve. Given some survival value from the ability to respond in a discriminating manner to certain molecules and ignore others, natural selection would then be a two-way process. In sexual reproduction, for example, if one sex happens to generate an unusual metabolic byproduct and if the other can perceive it, natural selection should operate with the eventual production of highly species-specific pheromones, kairomones, allomones, and other biologically important chemical signals. While these chemicals would tend to have more or less related constitutions in related species, there would be no across-the-board chemical similarity because their interactions are quantum-mechanical rather than chemical.

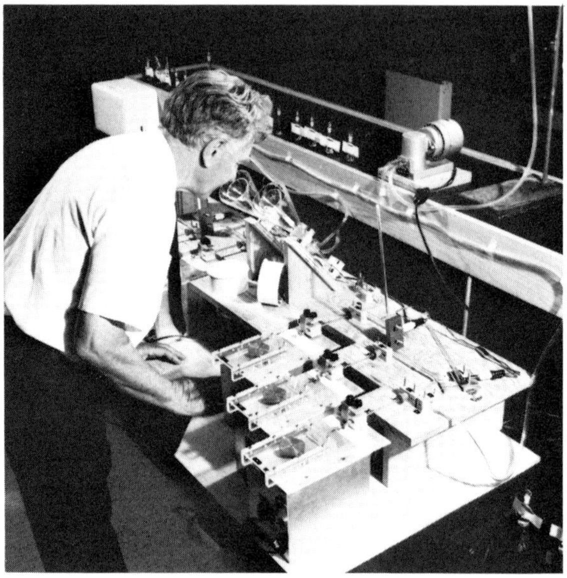

FIGURE 1. The apparatus used in an unsuccessful attempt to generate a blend with a musky odor. Electronically controlled syringe drivers injected 1, 2, or 3 different chemicals at known rates into the air stream leading to one smelling port. The smell of the blend could be compared with that of an authentic musk perfume at a second port.

Thus we find that insects employ the most varied sorts of chemicals as sex signals. The honeybee uses 9-oxo-2-decenoic acid; the bark bettle, *Dendroctonus frontalis,* uses a complex, bridged-ring compound, 1,5-dimethyl-6,8-dioxabicyclo [3.2.1] octane; the silkmoth uses bombycol, an unsaturated alcohol, (E).(Z)-10,12-hexadecadien-l-ol; and the housefly a hydrocarbon, (Z)-9-tricosene, and so on.

This is in striking contrast with the rather close similarity in the compounds which widely diverse organisms employ to carry out specific bodily functions of a purely chemical nature. Respiratory pigments are basically the same in the larger moths, the octopus, and in man, because they are serve the same function as transporters of oxygen. Hereditary characters are transmitted by the same sorts of DNA and RNA in oats and elephants. Cellulose, the structural material of plants, has interesting chemical analogies with chitin, the structural material of the arthropod cuticle.

Moreover, the source of the pheromones is not necessarily linked to the seemingly most relevant part of the organism. Sex pheromones do tend to be generated in glands which are associated with the reproductive organs, and this is only natural in view of the selective advantage of synchronizing the release of the signal with the maturation of the apparatus. However, this is not mandatory or universal: in the honeybee, the pheromone originates in the maxillary gland of the queen.

Starting with the simplest and most primitive organism, two equally simple sorts of mutation could give rise to generalist receptors on the one hand and specialists on the other. To augment a set of generalists, all that is necessary is to add one repeating unit to the receptor site so as to create an additional family of receptors, as explained in Chapter 12. To elaborate a specialist cell, the mutation must insert a second, a third, a fourth, . . . kind of site in the distal end of the cell. The larger the number the more specialized the cell. Yet with all this, the essential quantized interaction mechanism would remain unchanged.

The existence of odor-specialist units has so far been demonstrated only in insects where limitations of space and weight demand the utmost economy of material. As explained in the next chapter, the addition of a new type of generalist receptor makes a more than proportionate demand upon the Central Nervous System (CNS). In a specialist receptor the discriminating template is taken from the CNS and moved to the distal end of the system, thereby eliminating much of the central processing apparatus that would otherwise be needed.

It is by no means improbable that specialist receptors occur in vertebrates also. (Something of the sort was alluded to in Chapter 17.) Perfumers sometimes speak of a "pure odor", by which they apparently mean a sensation, produced by a single compound, which has not (as yet) been duplicated more or less approximately by mixing compounds with different odors. (An example of this phenomenon has been described by Kendall and Neilson[2] who mixed methyl salicylate and anethole in certain proportions to produce an odor like that of safrole and with no more than minor characteristics of methyl salicylate and anethole.)

The musk odor is commonly regarded as pure in this sense, and the vibrations of substances with that odorous character conform to a well-defined set of frequencies shown in Figure 2 Chapter 6. It is interesting, therefore, that several attempts I made to produce a musk-like sensation by mixing substances chosen to provide the correct combination of frequencies were completely unsuccessful. The apparatus is shown in Figure 1. It consisted of a blower pushing air through a capillary into which one, two, or three pure odorants could be injected at controlled rates by variable-speed syringe drivers and microliter syringes. A second, duplicate channel offered a comparison smelling point with an authentic musk perfume, namely a blend of Astratone, Versalide, and Musk Tibetene.

The failure of the experiment is explicable if musk is perceived via specialist receptors. With, say, 10^{15} odorous molecules per cubic centimeter the average distance between them is about 10^{-5} cm, or 1000 Å. Thus there would be rather a small probability of two or more molecules arriving at the cluster of sites on a given receptor cell at nearly the same time. Given appropriate diffusion times and distances, the difficulty does not arise if one molecule can present enough of the required frequency components to do the job.

Musk is an important messenger in the sexual life of several vertebrate species so that it would not be surprising if evolution had produced some specialized cells for the detection of a signal of such special significance. It is interesting, too, that J. Kloek[3] found that many persons used the words, "cedar-like" or "sandalwood" to describe the odors of various sexually related steroids. This led him to wonder whether this might be related to the use of these odors for incense.

The sweaty odor of certain fatty acids may also be perceived more by specialist than by generalist receptors. Amoore has described a genetically linked specific anosmia to this odor in a small percentage of the human population. The presence of several of these fatty acids in the vaginal secretions of some species of monkey suggests that their odors may have a biological significance, and certainly, dogs are known to be especially sensitive to them.

The existence of such specialist receptor cells would not in any way run counter to the general principles laid down in Chapter 2 and the general informational approach to olfactory phenomena. The possibility will, however, have to be kept in mind when the time comes to rewrite the preceding chapter of this book.

REFERENCES

1. **Boeckh, J., Kaissling, K. E., and Schneider, D.,** Insect olfactory receptors, *Cold Spring Harbor Symp. Quant. Biol.,* 30, 263, 1965.
2. **Kendall, D. A. and Neilson, A. J.,** Sensory and chromatographic analysis of mixtures formulated from pure odorants, *J. Food. Sci.,* 31, 268, 1966.
3. **Kloek, J.,** The smell of some steroid sex hormones and their metabolites, *Psychiat. Neurol. Neurochir.* 64, 309, 1961.
4. **Mayer, M. S. and McLaughlin, J. R.,** *An Annotated Compendium of Insect Sex Pheromones, Monograph Series No. 6,* Florida Agriculture Experiment Station, Institute of Food and Agricultural Sciences, University of Florida, Gainesville, August, 1975.
5. **Michael, R. P., Zumpe, D., Keverne, E. K., and Bonsall, R.W.,** Neuroendocrine factors in the control of primate behavior, *Rec. Prog. Horm. Res.,* 28, 665, 1972.
6. **Michael, R. P., Bonsall, R. W., and Warner, P.,** Human vaginal secretions: volatile fatty acid content, *Science,* 186, 1217, 1974.
7. **Neuhaus, W.,** Odor thresholds of dogs for ionone and ethyl mercaptan and their relation to other odor thresholds of dog and man, *Z. Naturforsch.,* 9b, 560, 1954.
8. **Preti, G. and Huggins, G. R.,** Cyclical changes in volatile acidic metabolites of human vaginal secretions and their relation to ovulation, *J. Chem. Ecol.,* 1, 36, 1975.
9. **Wright, R.H.,** Odor and molecular vibration: neural coding of olfactory information, *J. Theor. Biol.,* 64, 473, 1977.
10. **Wright, R. H.,** Specific anosmia: a clue to the olfactory code or to something much more important?, *Chem. Senses Flavor,* 3, 235, 1978.

Chapter 22

SOME ANATOMY

All of these endeavors are based on the belief that existence should have a completely harmonious structure. Today we have less ground than ever before for allowing ourselves to be forced away from this wonderful belief.

A. Einstein

By regarding the sense of smell as a channel of communication from the outside world into the organism, we have been forced from Chapter 1 onward to include in the same picture the observable properties of the stimulus and the observed responses of the organism. In the process we have had to postulate certain features of the transmission system — such as selectively tuned receivers — if the system is to have the capacity to do the things it is seen to do. By taking an overall view, we have tried to avoid the errors inseparable from a piecemeal approach. The time has now come to look at what is known about the actual construction of the mechanism.

Because olfaction involves stimuli of molecular dimensions, it is not surprising that many of the most interesting structures in the receptor apparatus are beyond the limit of even the electron microscope. Nevertheless, we can at least review the structures that are open to view.

The actual receptors are bipolar cells whose dendrites terminate in a small swelling from which project as many as ten or more olfactory cilia which may be as long as 100 μm (0.01 cm) but which, for most of their length, are no more than 10^{-6} cm in diameter (see Figure 1). In the rabbit there are about 140,000 receptors per square millimeter of receptor surface for a total of about 10^9 in the whole olfactory epithelium, so that the aggregate area of the cilia is very great: 10^9 cylinders 0.01 cm long and 10^{-6} cm in diameter have a total surface area of about 30 cm^2.

At its proximal end, each primary receptor is connected by its own nerve fiber which runs without interruption to the olfactory bulb, an arrangement that minimizes (but may not altogether eliminate) the opportunity for interplay between olfactory impulses before they reach the brain. Thus, in the main, each receptor cell must function as an independent unit through which impulses generated by interactions with stimulus molecules are conveyed directly into a forward extension of the cerebral hemispheres (the olfactory bulb) and with no synaptic interruptions on the way. Thus smelling is our most direct and intimate sensory contact with the outside world: odorous molecules impinge on the actual substance of nerve cells at one end, and the resulting signals arrive in the brain at the other end of the same nerve cells, with no intermediate structures to negotiate.

Within the olfactory bulb these nerves do not terminate in a diffuse arrangement but end in discrete formations, the olfactory glomeruli. There are about 1900 of these in the rabbit, each one being a spherical structure about 0.2 mm in diameter. In them the incoming and outgoing nerve fibers are intertwined with extraordinary intimacy and packed together like a very tightly wound ball of wool. There does not seem to be any systematic relation between particular glomeruli and particular areas of the olfactory epithelium, so that each glomerulus probably receives fibers from receptors which are widely dispersed across the receptive area in the nose. Within the glomeruli, the incoming axons from the receptors are in synaptic relationship with the dendrites of large mitral cells through which the signals pass onward to the Central Nervous System (CNS). In the rabbit, quantitative counts show that in each glomerulus, impulses from about 26,000 receptor cells are received and relayed onwards through 24 mitral cells.

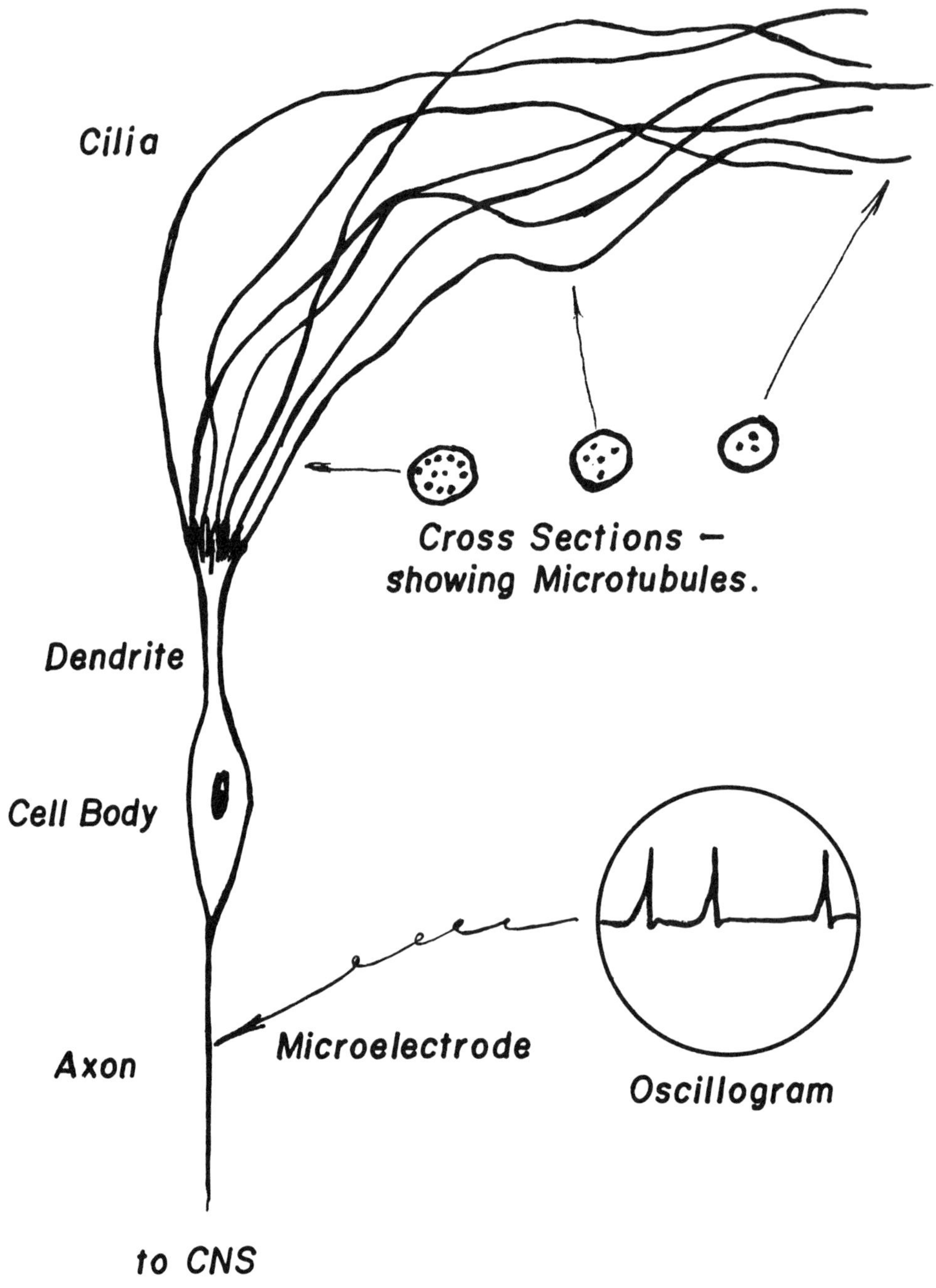

FIGURE 1. The outward end of a primary olfactory receptor nerve. The actual, molecular-size, receptor sites are located on the cilia which float in the mucus that bathes the olfactory epithelium and are in direct contact with the air when they lie in the surface of the film.

The number, 24, is interesting as suggesting the number of bits of information that suffices to ensure the rabbit's survival.

The axons from the receptor cells also make synaptic connections with so-called tufted cells to give a system of cross-linkages, somewhat as shown in Figure 2. The function of these inner structures and cross-links is almost entirely unknown, but Hai-

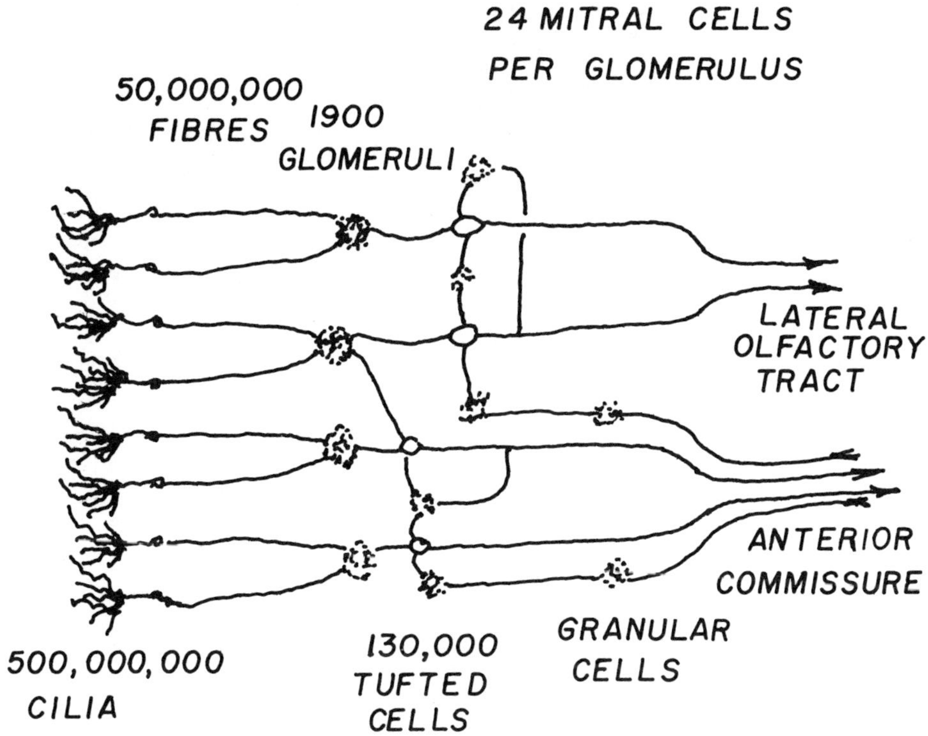

FIGURE 2. The signals generated by stimulus molecules interacting with olfactory hairs (cilia) are sorted out in the glomeruli so that all those coming from a particular type of receptor are channeled into a single mitral cell for passage onward to the brain. The quantitative values are those for the rabbit. The function of the tufted and granular cells and the various cross-connections can still only be surmised.

ner and his colleagues[6] have offered some interesting suggestions. For example, the funcion of the tufted cells may be to provide feedback loops by which the system could distinguish between genuine signals and random, spontaneous firing of the neurons. Furthermore, inasmuch as molecular impacts on the end organs are random in time, there may be a need for some timing or phasing device to hold the incoming signals briefly and synchronize their release though the mitral cells so that the image received by the CNS will not flicker. The granular cells (whose function appears to be unknown) may be well-placed to do this.

The figure shows efferent nerves coming from the CNS and hence having the potential ability to filter out parts of the incoming signal for special attention. One of the interesting things about smelling is our ability to direct our attention to various contributory notes in the overall sensation, notes to which we give names like, "spicy", "floral", "green", "balsamic", "putrid", and so on.

Thus there seems nothing in the known anatomical details of the apparatus as it exists in vertebrates that in any way rules out the picture we have developed of its operative mechanism.

In insects, notwithstanding their great structural differences, the actual nerve network is not very different, as can be seen from the following description, quoted from a paper by Boeckh et al.[4]

Insect's olfactory receptor cells . . . have dendrites ending in small pores on cuticle pegs, hairs and plates. Their neurites extend via the antennal nerve to the central organ where they are synaptically connected with the nerve cells of the deutocerebrum . . . A recent electron microscopical study . . . showed a good correlation between the number of small diameter nerve fibers in the antennal nerve of the silkmoth, *Bombyx*, with the number of antennal sense cells . . . The branched dendrites of cells innervating hairs and pegs end in pores of the cuticle which are approximately 150 Å wide. At this opening the receptor cell membrane seems to be in direct contact with the air . . .*

Thus the picture repeats that of the vertebrates: receptor cells dividing at their outward ends into extremely fine filaments which make direct contact with the outside, while at its inner end each receptor cell transmits signals via its own axon and with no synaptic interruptions between the external world and the processing centers of the CNS.

Evolutionists are by no means unfamiliar with examples of structures which have evolved to meet one need turning out to confer substantial degrees of survival value in quite unexpected directions. As mentioned in Chapter 2, the quantitative amount of information brought to the CNS doubles every time an additional, specifically sensitive generalist receptor is added to the system. Thus, every time a mutation increased the number of input channels it made a more than proportionate demand upon the information-processing resources of the CNS. It is in line with this that, as LeGros Clark[7] has put it, "The olfactory bulb, in which the olfactory nerves terminate, is developmentally an extension forwards of the cerebral hemisphere, and that the direct connection with it of the olfactory receptors is taken to be an expression of the fact that, from the evolutionary point of view, the cerebral hemispheres were initially developed in the vertebrate series in relation to the olfactory sense."

REFERENCES

1. **Adey, W. R.,** Higher olfactory centers, in *Taste and Smell in Vertebrates,* J. and A. Churchill, London, 1970, 357.
2. **Allison, A. C. and Warwick, T. T.,** Quantitative observations of the olfactory system of the rabbit, *Brain,* 72, 186, 1949.
3. **Andres, K. H.,** Anatomy and ultrastructure of the olfactory bulb in fish, amphibia, reptiles, birds, and mammals, in *Taste and Smell in Vertebrates,* J. and A. Churchill, London, 1970, 177.
4. **Boeckh, J., Kaissling, K.E., and Schneider, D.,** Insect olfactory receptors, *Cold Spring Harbor Symp. Quant. Biol.,* 30, 263, 1965.
5. **Gasser, H. S.,** Olfactory nerve fibres, *J. Gen. Physiol.,* 39, 473, 1956.
6. **Hainer, R. M., Emslie, A. G., and Jacobson, A.,** An information theory of olfaction, *Ann. N.Y. Acad. Sci.,* 58, Art. 2, 158, 1954.
7. **LeGros Clark, W. E.,** Inquiries into the anatomical basis of olfactory discrimination, *Proc. R. Soc. Ser. B.,* 146, 299, 1957.
8. **Ottoson, D.,** Some aspects of the function of the olfactory system, *Pharmacol. Rev.,* 15, 1, 1963.
9. **Reese, T. S. and Brightman, M. W.,** Olfactory surface and central olfactory connexions in some vertebrates, in *Taste and Smell in Vertebrates,* J. and A. Churchill, London, 1970, 177.
10. **Schneider, D.,** Specialized odor receptors of insects, in *Gustation and Olfaction,* Academic Press, London, 1971, 45.
11. **Slifer, E.,** The fine structure of insect sense organs, *Int. Rev. Cytol.,* 2, 125, 1961.

* From Boeckh, J., Kaissling, K. E., and Schneider, D., *Cold Spring Harbor Symp. Quant. Biol.,* 30, 263, 1965. With permission.

Chapter 23

PROBLEMS OF CLASSIFICATION

The knowledge of nature as it is — not as we imagine it to be — constitutes true science.

Paracelsus

A smell perceived is a message received. Chemical molecules are the medium, but the message is whatever interpretation is given to the signal by the recipient. The distinction between medium and message is best made plain by a variant of the philosophical problem of whether the moon continues to exist when no one is looking at it. Does a rose smell like a rose when there is no one there to smell it? If there is no one there, evidently there is no rose-like sensation being registered, but the molecules capable of stimulating that sensation are nevertheless there. The distinction between medium and message is vital when we approach the problem of classifying smells.

There is no problem with smells as external stimuli. We can classify smelly *substances* by chemical constitution, or solubility, or storage life, or price or . . . and one is as good as another if it lumps together things having a special character that interests the person doing the classification.

A different sort of problem appears when the object is to classify the messages. Smell sensations have been grouped into classes with names like ethereal, aromatic, fragrant, putrid, and so on. Such descriptive words can put the sensations into broad categories, but quite different sensations can be fragrant or nauseating. Moreover, many fragrances combine the odors of a variety of sources as in a perfume that is both floral and spicy.

To take this possibility into account, Crocker and Henderson selected four elementary odors which were thought to contribute in varying proportions (on a 1 to 8 scale) to the total sensory input. They were hoping to describe each odor by a 4-digit number so that, for example, the number 5532 meant a distinctly fragrant or flowery grade 5 note together with a 5 acid note also relatively strong and so implying a markedly fruity odor. The 3-rating for the burnt character meant a certain amount of woodiness, and the 2-rating for caprylic indicates a pleasant odor. "On assembling these impressions, we arrive at a moderately intense floral fruity odor". It would seem that the verbal description, fruity-rosy, is not more cumbersome than the odor number, and certainly the words convey more information than the digits 5532.

The compound with the 5532 rating was neryl acetate, and it is interesting to see how its odor was described by six professionals in the field of perfumery chemicals. They were asked to evaluate its odor for rose, green, and other characters with the following result.

Observer	Rose?	Green?	Other?
1	Medium	Weak	Fresh citrus
2	Pure	Part	Fruity
3	Strong	–	Oysters (!)
4	Yes	None	Fruity
5	+		Fruity
6	Medium	Slight	

One gets the impression that there is more to the message than can be spelled out in a 4-digit number.

For these and other reasons, the earlier attempts to classify odors faded away with an obituary notice by E. Sagarin with the interesting title, "On the Inherent Invalidity of All Current Systems of Odor Classification."[2]

More recently, multi-dimensional scaling of odor similarity or dissimilarity has been invoked as a basis for classification. Thus, for example, observers have been asked to rate pairs of stimuli on the basis of pleasantness/unpleasantness, but as M. H. Woskow has pointed out, ". . . we are soon confronted with the embarrassing finding that some odors appear equally pleasant but are not at all alike".[3] He then goes on, ". . . we do not know the dimensionality of odor space. Yet we do know that some odors are alike and others are dissimilar, and that these relationships should be indicated in the odor space."[3]

Accordingly, he undertook a study which disregarded the relative hedonic values of two odors and concentrated simply on their estimated similarity/dissimilarity as scaled subjectively in such a way that category judgments could be treated as absolute distances. Experimentally, twenty adult observers were asked to estimate the total similarity (rated 1) and the total dissimiliarity (rated 9), or intervening separations, on pairs of odorants drawn from a set of 25 compounds. This gave some 300 paired comparisons by each of 20 observers. Thus, for example, it was found that the mean estimate of the distance between methyl salicylate and eugenol was 2.60 units on the 1 to 9 scale. With a nine-unit scale, it was possible to assign each odorant a position in nine-dimensional space, but with some residual scatter which became more pronounced when the separations were fitted into seven, five, and three dimensions. Generalizing on this result, one might say that it would take as many dimensions as there are compounds to assign each odor a unique position in a many-dimensional hyperspace.

But is this not merely another way of saying that every substance has an odor that is unique to it — a conclusion strongly supported by matters presented in Chapter 2.

Multi-dimensional scaling is nevertheless valuable as demonstrating the dimensionality of the problem. It is analogous to a measurement of the distances separating any two of the following: London, the Panama Canal, Cape Horn, the Lake of Geneva, the North Pole, and Mount Everest. The distances can be reliably determined, but would tell us little about the geography of the world because of the dissimilarity in the nature of the fixed points. What the exercise would clearly establish is the shape of the surface of the earth, when it is found that the distances cannot be plotted on a flat, two-dimensional piece of paper, but can be so plotted on the surface of a sphere.

All these difficulties bring us back to a consideration of what a system of classification is and does. Classifications are manmade for man's convenience and a classification based on pleasantness/unpleasantness can well vary with the situation in which the estimate is made. The pleasantness or otherwise of a food odor will depend upon whether it is smelled in a dining room or a public library, and that of a perfume will vary as between a bank and a ballroom.

In every case, a smell perceived is a message received, but the information in the message will be related to the particular situation of the receiver.

For purposes of understanding the mechanics of smell perception, we need a system of classification that will lead to an unequivocal insight into the one-to-one relation that must exist between particular molecular attributes in the stimulus and particular nervous discharges in the organism. Such a classification is essential to understanding sensory mechanisms but would be almost useless in estimating the quality of a wine or the amount of nuisance from a factory making fertilizer.

The recognition that smells are messages suggests an analogy with words. How do we classify words, and can our classification of them provide suggestions for classifying smells?

Grammarians have classified words on the basis of their function: as nouns, pronouns, adjectives, verbs, adverbs, prepositions, conjunctions, and interjections. But since, "and", "or", "also", "moreover", "besides", etc. are all conjunctions, there is evidently little to be learned of their meaning from the mere fact that they are all conjunctions. To determine the meaning we have to turn to the dictionary, and there we find an interesting clue. Consider the following definitions of the word "horse":

1. A solidungulate, perissodactyl mammal of the family Equidae and the genus *Equus*[4]
2. A solid-hoofed quadruped with a flowing mane and tail, used as a beast of burden and draught and for riding on[5]

By itself, the first definition would convey little real information *were it not accompanied by a picture of a horse.* The second definition replaces the picture with the verbal description of the flowing mane and tail and a mention of the sort of context in which a horse is likely to be encountered.

Each definition, then, seeks to give the meaning of the word in relation to the sort of visual inputs on which most of us would rely for information about the outside world. An interesting exception is the way one dictionary defines the word "goat": a hardy, lively, wanton, strong-smelling, usually horned and bearded ruminant quadruped.

Where smell sensations are concerned, we can approximate a scientifically useful classification system by pursuing the dictionary analogy. We set about assembling an unlimited number of standard samples which will be given names like musk, rose, peppermint, garlic, sandalwood, lemon, bitter-almond, green, sweaty, fishy, musty, catty, and so on *without limit.* A Dictionary of Smells is then a row of bottles each with a word on the outside and the equivalent of a picture on the inside. And just as a donkey has some but not all the attributes of a horse, so a given compound may have some, but not all the attributes of any one standard, and will require several words to describe its odor.

The solution to the problem is to discard the word "classify" and replace it with the word "categorize". A compound with an odor described as weak almond with a trace of cumin alongside sassafras can only be classified as having a complex odor. It is much more useful to categorize it as having some bitter-almond, some cumin, and some sassafras components in its overall message, and anyone unfamiliar with the odor of sassafras can find the meaning of the word in the Odor Dictionary.

The compounds used in developing the peak number plots in Chapter 6 were assembled on the basis of their having been categorized by qualified persons who used words in their dictionary meanings.

The advantage of this dictionary approach is that a compound which manifests a complex odor (needing several words to describe it) will be especially useful in enabling us to trace its varied sensory inputs back to various subpatterns in the overall attribute pattern of its molecules in the way touched upon in Chapter 8.

It is because classification is a means to an end rather than an end in itself that it has been possible to defer any discussion of the matter until this stage in my account, and to dismiss the word "classify" in favor of the more appropriate word, "categorize".

REFERENCES

1. **Crocker, E. C. and Dillon, F. N.**, Odor directory, *Am. Perfum. Essent. Oil Rev.*, 53, 297, 1949.
2. **Sagarin, E.**, On the inherent invalidity of all current systems of odor classification, *J. Soc. Cosmet. Chem.*, 2(1), 25, 1950.

3. **Woskow, M. H.,** Multi-dimensional scaling of odors, in *Theories of Odor and Odor Measurement,* Tanyolac, N. M., Ed., Robert College Research Center, Bebek, Istanbul, 1968, 147.
4. *The Century Dictionary,* The Century Company, New York, 1914.
5. *The Concise Oxford Dictionary,* The Clarendon Press, Oxford, 1934.

Chapter 24

WHAT IS NORMAL?

We have just used the words "abnormal" and "anomalous" . . . The words simply mean that in groping for the truth, an unexpected result has been obtained, which once stood, or now stands, challenging investigators to show how the unexpected should have been expected.

J. W. Mellor

In Chapter 6, some of the difficulties of working with human judges of odor were mentioned. The time has come to examine them more closely in whatever light theory can shed. Consider, first, the simple matter of sensitivity.

Olfactory sensitivity varies with the individual so that in a study of 30 subjects' threshold of perception for dimethyl sulfide, the values ranged from 50 parts per thousand million of air down to 3, giving a ratio of 16:1. This may be due to age — older people tend to be less sensitive than young ones — or to the nutritional state — thresholds are generally lower before a meal than after it — or to the effects of practice and training, but even when these are taken into account there still remains much innate variation. Fortunately, for clinical cases, the pathological findings are usually pronounced enough to leave no doubt of a really abnormal variation.

One well-documented peculiarity is the complete, or nearly complete inability of about one fifth of the population to smell the bitter-almond odor of hydrocyanic acid. It has been attributed to a sex-linked recessive gene, but does not seem to be a simple, genetically controlled trait. Such specific anosmias are not uncommon. For example, J. E. Amoore[2] found that approximately 2% of the population have difficulty in perceiving the sweaty odor of isovaleric acid, and certain other fatty acids. He has also described an apparently genetically based inability in others to smell the musky odor of pentadecanolactone. It is interesting that both these odors are biologically important and may therefore be perceived more through the medium of specialist than generalist receptors (Chapter 21), in which case they may not provide as much insight into the olfactory code as has been hoped.

Some specific anosmias are a little surprising, not only because of their rarity but also because they have so little apparent relation to the survival prospects of the person exhibiting them. Thus, in a survey of 4030 persons, 4 were discovered who could not smell butyl mercaptan which is a principal component of skunk odor, even though they had no trouble with such odors as vanilla, oil of orange, or hydrogen sulfide. A somewhat similar situation turned up when 80 students were asked to smell each of two varieties of freesias and say whether they were or were not fragrant. Most found both varieties fragrant to some extent, but about one sixth of the group said that at least one species had no odor.

Apart from the possibility that such effects are the result of genetic or other abnormalities in the actual machinery, the responses to various odorous stimuli can be modified by medication or physiological condition.

In 1949, Guillot and Fiehrer[9] reported that to normal women, pentadecanolactone seems to have a strong and often disagreeable odor between menstrual periods, and a very weak, pleasant odor during menstruation. (This is in addition to the specific anosmia to this substance reported by Amoore.[2]) To normal men and pregnant women it had little or no odor. Commercially, this chemical is sometimes called Exaltolide, and the phenomenon is referred to as the "Exaltolide Phenomenon". A number of later reports appear to support the reality of the phenomenon, although my own experiments do not support the claim that women find the smell unpleasant. When asked to describe it they often use words like "perfume" or "flowers", but they do appear to detect the odor more reliably than men.

The effects of prior medication are at least as striking and probably much more common sources of confusion.

LeMagnen[16] injected odorous materials subcutaneously and showed that after an initial period of a few days during which the threshold was raised, there followed a period of abnormal sensitivity. The effect was specific to the odorant injected. Somewhat similar effects have been found with dogs following the oral administration of butyric acid or ionone. A less specific type of effect was observed with workers in a factory making menthol who showed consistently higher threshold scores than a control group. Menthol has some local anaesthetic properties which probably contribute to its use as a commercial air freshener.

Interesting effects have been reported following the administration of various psychoactive drugs to subjects who were healthy rhinologically and psychoneurologically. With 12 of 15 subjects for example, the threshold was lowered by Imipramine HCl which stimulates the rhinencephalon and it was also lowered by a meprobamate which depresses the rhinencephalon. Imipramine also altered some of the subjective interpretations of the odors: disagreeable ones stayed disagreeable, but finer ones, such as essence of rose, evoked sensations of nausea and disgust to the point of almost inducing vomiting in at least half the subjects.

The examples so far described are mainly specific anosmias of a permanent sort, or temporary aberrations with a (probably) ascertainable cause. Sometimes we find people who can perceive an odor, but apparently the odor they perceive is qualitatively different from that perceived by the majority. The term "parosmia" has been applied to this type of abnormality which is usually revealed by the use of an unusual word to describe the odor. Thus, Guillot[10] was able to distinguish three types of parosmia to the compound androstanone-3 from an examination of 50 persons. Some of them termed the odor extremely violent, foul, and sometimes even unbearable. Others compared it to a light musk, extremely pleasant and flowery. Still others said it was fruity, something like peach.

When I tested 60 people with 5 standard odorants I found another example. Whereas most people easily and reliably recognized the odor of peppermint, judging by the words they used to describe it, one man consistently used the words "musty-earthy". This is similar to the case of a woman mentioned by Guillot[10] who could not smell the violet odor of ionone but could detect its woody note.

This kind of abnormality has too often been put aside as inexplicable and thereafter ignored. If the vibrational frequency-patterns that correlate with, say, the violet and woody sensations were known (unfortunately as yet they are not), it might be possible to predict other parosmias the subject would manifest, and eventually to sort out the details of the code so as to make the inexplicable explicable. The enquiry would not be entirely academic or theoretical. For example, it is known that olfactory sensitivity can be diminished by prior exposure to subliminal stimuli. This is something that may have to be taken into account when water is polluted from several sources.

With this sentence our journey through Osphresiology — the Science for All Seasons — has moved from structural organic chemistry and quantum physics by way of electrophysiology and neuroanatomy into the mysterious world of the mind, and finally out into the workaday world of practical application. When Benjamin Franklin was asked what was the use of flying his kite into a thunderstorm he is said to have asked in return, "What use is a baby?"; and when Michael Faraday was asked the same question about his discovery of electromagnetic induction, his answer was equally apposite, "Someday you will be able to tax it!"

What, then, are the revenues to be expected from this baby of ours?

REFERENCES

1. **Amirov, R. Z.,** Effect of sub-threshold olfactory stimuli on the sensitivity of the sense of smell in normal and pathological conditions (transl.) *Biol. Eksperim. Bio. Med.,* 47(5), 562, 1959.
2. **Amoore, J. E.,** Specific anosmia: a clue to the olfactory code, *Nature (London),* 214, 1095, 1967.
3. **Amoore, J. E., Venstrom, D., and Davis, A. R.,** Measurement of specific anosmia, *Percept. Motor Skills,* 26, 143, 1968.
4. **Amoore, J. E., Popplewell, J. R., and Whissell-Beuchy, D.,** Sensitivity of women to musk odor: no menstrual variation, *J. Chem. Ecol.,* 1, 291, 1975.
5. **Blakeslee, A. F.,** Are freesias fragrant? *Natl. Hortic. Mag.,* 11, 211, 1932.
6. **Brown, K. S. and Robinette, R. R,** No simple pattern of inheritance in ability to smell solutions of cyanide, *Nature (London),* 215, 406, 1967.
7. **Cluzel, J. C. J.,** Contributions of the Olfactometer to the Studies of Some Endocrine Syndromes, thesis, Marseilles, 1964.
8. **Guillot, M.,** "Some characteristics of the phenomenon of partial anosmia, *C. R. Soc. Biol. Paris,* 142, 161, 1948.
9. **Guillot, M. and Fiehrer, A.,** Folliculin and olfactory sensation, *C. R. Soc. Biol. Paris,* 143, 922, 1949.
10. **Guillot, M.,** Anosmia and parosmia: a few individual cases, *Recherches,* 3, 26, 1953.
11. **Keith, E. S. and Powers, J. J.,** Determination of flavor threshold levels and sub-threshold, additive and concentration effects, *J. Food. Sci.,* 33, 213, 1968.
12. **Kendall, D. A. and Neilson, A. J.,** Correlation of subjective and objective odor responses, *Ann. N.Y. Acad. Sci.,* 116, Art. 2, 567, 1964.
13. **Kirk, R. L. and Stenhouse, N. S.,** Ability to smell solutions of potassium cyanide, *Nature (London),* 171, 698, 1953.
14. **Köster, E. P.,** Olfactory sensitivity and the menstrual cycle, *Int. Rhinol.,* 3(1), 57, 1965.
15. **Köster, E. P.,** Sex differences in odor perception, *Cosmet. Perfum.,* 90(5), 50, 1975.
16. **LeMagnen, J.,** Study of the phenomenon of olfactory sensitization, *C. R. Acad. Sci., Paris,* 228, 122, 1949.
17. **LeMagnen, J.,** Olfactory sexual phenomena in man, *Arch. Sci. Physiol.,* 6, 125, 1952.
18. **Naus, A.,** Alterations of smell acuity caused by menthol, *J. Laryngol. Otolaryngol.,* 82, 1009, 1968.
19. **Neuhaus, W.,** Variation of the olfactory acuteness of the dog by oral administration of odoriferous substances. *Z. Vgl. Physiol.,* 41, 221, 1958.
20. **Patterson, P. M.and Lauder, B. A.,** The incidence and probable inheritance of "smell blindness" to normal butyl mercaptan, *J. Heredity,* 39, 295, 1948.
21. **Semeria, C., Sartoris, A., and Busca, P.,** Psychotrop medicines and olfactive function, *Rev. Larnygol. Otol, Rhinol., Bordeaux,* 86th, Year Supp. Oct., 1965, 931—934.
22. **Vierling, J. S. and Rock, J.,** Variations in the olfactory sensitivity to exaltolide during the menstrual cycle, *J. Appl. Physiol.,* 22, 311, 1967.
23. **Whissel-Beuchy, D. and Amoore, J. E.,** Odor-blindness to musk: simple recessive inheritance, *Nature (London),* 242, 271, 1973.
24. **Zilstorff-Pederson, K.,** determination and variations of olfactory thresholds, *Arch. Otolaryngol.,* 79, 412, 1964.

PART II
Applications

> These investigations of a rather theoretical sort are capable of much more immediate practical application than one would be inclined to believe.
>
> H. L. LeChatelier

In 1950 the world's output of scientific literature included perhaps 100 papers related to smells and smelling. In 1973, *Chemoreception Abstracts* listed 1294 titles and by 1977 the number has risen to 1562. This dramatic increase reflects the long-delayed recognition that the chemical senses are not merely interesting to study, but have wide-ranging practical applications, not least of which is the control of insect pests without recourse to environmentally hazardous pesticides.

In this section, I will look at a few of these practical applications, mainly for the purpose of showing their diversity and some of the openings presented by the newer theoretical insights. A comprehensive review of the entire subject would have required my spending so much time reading that this book would not and could not have been begun.

Chapter 25

SUBSTITUTES FOR A SILENT SPRING

The most alarming of all man's assaults upon the environment is the contamination of air, earth, rivers and sea with dangerous and even lethal chemicals.

Rachel Carson

Man is everywhere beset by insects who attack his body directly when they cause disease or indirectly when they destroy his food. The record of victories against these enemies is impressive. It ranges from massive engineering works to drain the swamps where malarial mosquitoes breed through insecticides like lead arsenate or Paris green to DDT and the chemisterilants. The very success of these measures makes the residue of unconquered enemies all the more conspicuous, and even our most spectacular victories have a habit of turning out to be less than total when an organism counterattacks by becoming resistant to our arsenal of medicines and pesticides.

Evidently, no single defense strategy can be effective against all our insect enemies, and therefore every possible weapon must be explored and exploited in whatever way will give the maximum return for the minimum outlay. To this end, more and more attention is being given to a type of pest control using nontoxic behavior-modifying chemicals, or metarchons, which may be expected to supplement some conventional pesticides and replace others. Pesticide pollution of the environment is a real and present problem, but it is surely more sensible to make the pollution obsolete than to make it illegal.

Following are two examples of what may be achieved.

On the morning of November 1, 1962, a Grumman amphibian airplane flew over the 33 mi^2 island of Rota, 37 miles north of Guam. It carried pieces of fiberboard about 4 cm square which had been soaked in a chemical called methyl eugenol plus 3% of the insect poison Naled. When the plane landed, 125 of the fiber squares had been dropped on each square mile of the island except in the villages where they were hung in the trees by hand. The process was repeated 15 times in the next 6 months.

For 30 years the island had been plagued by an insect, the oriental fruit fly, *Dacus dorsalis,* which infested breadfruit, papayas, and mangoes and which had resisted every attempt to control it. Following the tenth drop and for more than 4 years thereafter, not a single male fruit fly was found anywhere on the island. With no males to mate with, the females produced no fertile eggs and that ended the matter.

- The pest was not controlled — it was eradicated.
- The total amount of Naled used was 1/10 oz/acre.
- The whole operation cost less than $1/acre.
- No other insect was directly affected.

The story began to take shape in 1952 when L. F. Steiner,[5,6,7] working with the U.S. Department of Agriculture in Hawaii, tested various chemicals to see whether they would attract oriental fruit flies. He found that young males were attracted to methyl eugenol and would eat it with gusto even when it was spiked with a poison like Naled. On Rota, when the first fiberboard squares were dropped, males were seen to fly upwind from considerable distances to get to them, and the squares were still attractive 2 months later.

Total success was achieved on Rota and subsequently in Saipan and Tinian because a whole island could be treated and reinfestation from outside was difficult because

Table 1
***TRICHOPLUSIA NI:* DISRUPTING SEX PHEROMONE COMMUNICATION WITH LOOPLURE[a]**

Area treated	144 ha = 355 acres
Distance between evaporators	200 m
Total looplure per night	3.6 = 0.107 oz
Males per trap-night, treated area	1.2 average
Males per trap-night, control area	124.7 average
Average disruption of mating	99.3%

[a] *cis*-7-dodecenyl acetate

of the open water. On the larger islands in Hawaii, only cultivated areas could be treated, but similar infestations were reduced by 80%. They were not wholly ended because fertile females can live for as long as 3 months and move into areas where the treatment has left no living males. Yet even without complete eradication, the pest was kept under control at a cost of less than 25 ¢ per acre per year.

This is an example of the male annihilation technique of pest control.

The next example illustrates the method of male confusion.

Looplure, or *cis*-7-dodecenyl acetate, is the synthetic sex pheromone of the cabbage looper moth, *Trichoplusia ni.* Professor Shorey[3] and his colleagues at the University of California in Riverside released it into the air over large experimental plots in the Coachella Valley in California from evaporators arranged in a 6 × 6 checkerboard with the evaporators spaced 200 or 400 m apart. The evaporators were designed so as to release the looplure at a controlled and measured rate. It was known that when male moths were exposed continuously or intermittently to the synthetic pheromone they found it difficult to respond to the emanation naturally released by the females, and the object of the experiment was to find out whether, and how much, wild males would be frustrated in their attempts to find mates by the broadcast release of the chemical. To test its effectiveness, a trap containing virgin female moths 3 to 7 days old was placed in the center of each field plot and the number of male moths captured was recorded each day. The results are summarized in Table 1.

It is clear that the insects can be almost wholly prevented from mating by the release of almost infinitesimal quantities of the pheromone: 3.6 g per night spread over an area of 144 ha, or, in possibly more familiar units, 0.0003 oz. per acre per night!

The choice whether to use a few dispensers widely dispersed and each releasing a relatively large amount of chemical or many relatively weak sources closer together is partly a matter of economics and partly determined by the behavior of the target insect. There are indications that disruption of pheromone communication in the pink bollworm, *Pectinophora gossypiella,* breaks down when evaporators containing Gossyplure (the sex pheromone of this species) are more than 60 m apart. Other factors which may affect the outcome are, naturally, weather, terrain, and the types of vegetation present.

As was the case in Rota, in these experiments, only one species of insect was directly affected, and here no toxic material was employed. The cabbage loopers were controlled by modifying their behavior by preventing them from receiving the signals on which they normally depend to manage their affairs.

The male annihilation and male confusion techniques are only two of the possible ways behavior-modifying chemicals or metarchons might be used to control insect pests without leaving hazardous residues in the environment and without harming any other plant or animal species. Females, for example, usually lay their eggs in places where the young are likely to find their proper food. If they can be induced to lay them in the wrong place they will be controlled.

A pest is only a pest when it pesters.

In principle, every kind of insect behavior can be modified to the detriment of the insect and the benefit of man.

Controlling an insect by changing its behavior is a very different thing from controlling it with a poison. A poison is a molecular monkey wrench thrown into the chemical machinery an insect uses to grow, or make eggs, or move its legs, or flap its wings. Some individuals are less easily damaged in this way than others, and the resistant ones tend to pass on their resistance to their offspring, so that sooner or later every species becomes resistant to each new pesticide. Contrast this with the machinery an insect uses to regulate its behavior. It receives information from its eyes and other sense organs, and on the basis of these signals its brain sends out appropriate orders to walk, fly, eat, mate, lay eggs, or whatever else is needful for the survival of that species at that time and in that place.

Since most if not all insects have room for only a very small brain in their rather small bodies, their responses have to be almost entirely automatic. Under natural conditions, if a male receives the sensory input from the female scent it can have come only from a female and so his response is as appropriate as it is automatic. If we present him with the same signal in the absence of a female, he will begin by responding inappropriately and end by ignoring a real female when he meets one.

REFERENCES

1. **Farkas, S. R., Shorey, H. H., and Gaston, L. K.,** Sex pheromones of Lepidoptera. The use of widely separated evaporators of looplure for the disruption of pheromone communication in *Trichoplusia ni, Environ. Entomol.,* 3, 876, 1974.
2. **Farkas, S. R., Shorey, H. H., and Gaston, L. K.,** Sex pheromones of Lepidoptera. The influence of prolonged exposure to pheromone on the behavior of males of *Trichopluia ni, Environ. Entomol.,* 4, 737, 1975.
3. **Shorey, H. H.,** Use of pheromones in pest control, *Proc. N. Cent. Branch Entomol. Soc. Am.,* 27, 30, 1972.
4. **Shorey, H. H., Kaae, R.S., and Gaston, L. K.,** Sex pheromones of Lepidoptera. Development of a method of pheromonal control of *Pectinophora gossypiella* in cotton, *J. Econ. Entomol.,* 67, 347, 1974.
5. **Steiner, L. F.,** Methyl eugenol as an attractant for oriental fruit fly, *J. Econ. Entomol.,* 45, 241, 1952.
6. **Steiner, L. F., Mitchell, W. C., Harris, E. J., Kozuma, T. T., and Fujimoto, M. S.,** Oriental fruit fly eradication by male annihilation, *J. Econ Entomol.,* 58, 961, 1965.
7. **Steiner, L. F., Hart, W. G., Harris, E. J., Cunningham, R. T., Chinata, K., and Kamakahi, D. C.,** Eradication of the oriental fruit fly from the Mariana Islands by the methods of male annihilation and sterile insect release, *J. Econ. Entomol.,* 59, 131, 1970.
8. **Wright, R. H.,** "Metarchon" a new term for a class of non-toxic pest control agents, *Nature (London),* 204, 603, 1964.
9. **Wright, R. H.,** Some alternatives to insecticides, *Pest. Sci.,* 1, 24, 1970.

Chapter 26

SUBSTITUTES FOR A SILENT SPRING — II

The concepts and practices of applied entomology for the most part date from the Stone Age of science.

Rachel Carson

The matters presented in the preceding chapter are so intimately relevant to the problems of our time that, at the risk of becoming repetitious, I will recapitulate in generalities what has been said in particular.

When we speak of controlling an insect, what we usually mean is control of its numbers. This may enable us to minimize the damage it does, but controlling the size of the population is not the same as controlling the insects themselves. To control an insect in any real sense, we must learn to order it about — and to have our orders obeyed! To exercise this kind of control, we must know and use the language that the insects understand and obey. This is a field of investigation that is only now beginning to be explored. It is one that may someday turn out to have surprising byproducts, some of which will be discussed later in Chapter 48.

Insect language is mostly a series of categorical imperatives: "Do this!" or occasionally, "Don't do that!" with the understanding that the order is to be obeyed only as long as it is reiterated. When the order is stopped or changed the response stops or is changed. Insect brains are mostly too small to permit much independent thought and the apparently concerted and purposeful activities they display are the result of an essentially blind obedience to orders. For example, we can direct a mosquito to alight on a designated spot by causing the air currents emanating from the spot to have the correct combination of temperature and humidity. The language in which we give the order is here a physicochemical one because it involves both the chemical composition of the air stream and its physical properties of temperature and movement.

Movement is very important because smell by itself is nondirectional. We found that the way fruit flies, *Drosophila melanogaster,* react to the smell of fermenting banana depends upon whether the air carrying the scent is moving or still. If the air is uniformly scented and is moving, the insects will fly steadily upwind; and if it is uniformly scented and not moving they alight and crawl about in all directions. If the air is moving but not uniformly scented, then the scent reaches the insect as an irregular series of sensory inputs. Each time it perceives the scent it starts upwind, and each time it loses the scent it turns across the wind and may later go back downwind. The resulting flight pattern looks like a highly purposeful backtracking of the scent trail, but in fact is a pattern of mechanical responses to a very limited number of orders, some of which are given chemically but others of equal importance may be given mechanically, or thermally, or optically.

The insect's Central Nervous System (CNS) processes the words of command according to a set of rules of grammar before issuing the appropriate messages to its motor neurons, which is why these studies have such potential as affording insights and clues to mental operations generally.

Consider the matter of air movement or lack of movement in relation to the response of a flying insect to an attractive scent. A creature standing on the ground or clinging to a tree or a blade of grass can know which direction the wind is coming from by the way its fur or whiskers or antennae are bent with respect to the rest of its body, which cannot yield to the breeze as long as it clings to a surface. The situation becomes entirely different as soon as it takes wings and flies: now it will move with the air but not through it unless its own swimming movements generate a relative movement. The only way it can purposefully direct its flight into or across or with the wind is by using its eyes to register its movement with respect to objects nearby.

To demonstrate this, we put insects in a wind tunnel with a transparent floor onto which we could project moving spots. If the spots were not moving and the air was, the insects easily made their way to the upwind end of the tunnel or to the tube through which an attractive scent was being released. To do this, they had to align their bodies so that when a spot came in view it passed under the long axis of the insect relatively slowly from front to rear. If the insect happened to be flying with the wind, the spot would pass under it relatively quickly. Thus, the central processing mechanism must include a time sense! By moving the pattern of spots on the floor of the tunnel at an angle to the wind, we could steer the insects past the source of scent, and by moving the spots in the same direction as the wind and a little faster than the wind movement, we could make them turn around and fly with the wind and *away* from the source of the scent they were seeking!

This interpretation of the behavior of insects in their search for the source of an odor is evidently more realistic than the alternative which has sometimes been suggested, namely, that they fly up a gradient of increasing concentration. One has only to look at the smoke from a factory chimney or a cigarette to realize that at any one instant the pattern is very different from the mathematically formulated picture and calculated average. Furthermore, even if we take the mathematical diffusion gradient at its face value, the insect might have to fly tens or even hundreds of meters before it could register a Just Noticeable Difference (JND) in the strength of the smell. This is true whether the Fechner-Weber or the Stevens method of relating sensation to stimulus is adopted. Experiments show that these psychophysical laws are, indeed, applicable to insects.

Given, then, that insect behavior is governed by a language which has both words and grammar, we begin to view the examples described in the last chapter in a new way. When the cane-fiber squares impregnated with attractive scent and poison were dropped on the islands, each square acted as a point source which the male flies could locate by the means described above. When Professor Shorey and his group permeated an entire area with female scent this was no longer possible. What seemed to be well-nigh impossible was the amazingly small amount of chemical needed to do the job. The following calculation shows how it is possible.

The sense of smell is so sensitive that it has been shown that the male American cockroach, *Periplaneta americana,* will react to fewer than 30 molecules of the female's attractant scent. Suppose we have a chemical which is biologically active at a concentration of, say, 1000 molecules per cubic centimeter. A concentration 100,000 times this, or 100 million molecules per cubic centimeter should be enough to saturate the receptors of a male insect. But 1 gram molecule of the chemical (in other words, a few ounces) contains 6×10^{23} molecules, so that a pound of it should overwhelm the normal responses when dispersed in 6 mi^3 of air!

It remains only to mention a way of exploiting insect language which has so far not been mentioned. Besides leading the insect to the wrong goal or blinding it to the existence of any goal, we can turn it aside by causing the orders to be received in the wrong sequence, or, figuratively, by producing mistakes in the grammar. This is what happens when we apply a chemical repellent to our hands and faces to keep off mosquitoes and other biting flies. These so-called repellants are not merely attractants with a minus sign. If they were they would send the mosquitoes as far downwind as our bodily attraction normally brings them upwind. The sequence of actions and reactions the mosquitoe employs to find its way to a source of blood, requires it to turn at certain places and not turn at others. The effect of the repellent molecules released along with our normal bodily emanations is to put in turns where there should be no-turns, and no-turns where there should be turns.

Repellency and attractancy are not properties of particular chemicals but are, rather, end results of these chemicals putting sensory inputs into an insect's CNS which it then coordinates and interprets and eventually manifests a response. It may be convenient to call a certain chemical the sex attractant of a certain insect, but evidently sex attraction involves much more than the sex-attractant chemical.

This fact was brought home to me very vividly in the following way. Professor J. H. Borden[1] at the Simon Fraser University isolated and identified the aggregation pheromone of the ambrosia beetle, *Gnathotrichus sulcatus.* It is a chemical called 6-methyl-5-hepten-2-ol, or sulcatol. He gave me a small sample for spectroscopic examination and I found that it had many frequencies approximately matching some of the frequencies of menthol. A few days later, Professor Borden telephoned to say that in a laboratory bioassay, menthol was nearly as effective as sulcatol. This was eminently gratifying, but some days later he called to report that in a field-test menthol showed no attraction whatsoever.

The difference was that the laboratory test was a walking bioassay and in the field the beetles had first to fly to the vicinity of the attracting source and walk the rest of the way. Evidently, menthol gives the walking orders but not the flight plan. Later, Professor Borden found that the natural pheromone is a mixture of two mirror-image isomers and the insects responds only when both isomers are present. As explained in Chapter 15, even though mirror-image molecules have the same frequency-patterns most of the time, when they are near another asymmetric structure they are no longer mirror-images and can therefore convey different messages to the organism.

A. A. Sekul et al.[4] had a similar experience with the corn earworm moth, *Heliothis zea.* Its sex attractant was isolated and identified as (Z)-11-hexadecenal and was seen to elicit strong sex stimulation reactions in caged laboratory males. However, just as with menthol, the chemical did not catch males when used to bait traps in corn fields. Their experiments were then carried one step farther when minute quantities of the chemical were included in traps baited with as many as four virgin female moths, and their normal allure was completely inhibited.

The question is, however, whether "inhibition" is in fact the correct word.

If attraction is the end result of a series of orders being given and of responses being elicited *in the correct sequence,* and if the correct sequence is something like "fly now, walk later", there will be no observable aggregation under natural conditions if the order "Walk!" drowns out the instruction to "Fly!" It is not enough to label something with a word like "inhibition" if it is in fact a consequence of orders arriving in the wrong order or with the wrong emphasis.

REFERENCES

1. **Borden, J. H., Chong, L., McLean, J. A., Slessor, K. N., and Mori, K.,** *Gnathotrichus sulcatus:* synergistic response to the enantiomers of the aggregation pheromone, sulcatol, *Science,* 192, 894, 1976.
2. **Kennedy, J. S.,** The visual responses of flying mosquitoes, *Proc. Zool. Soc. Ser. A,* 109, 221, 1939.
3. **Kellogg, F. E., Frizel, D. E., and Wright, R. H.,** The olfactory guidance of flying insects. IV. *Drosophila, Can. Ent.,* 94, 684, 1962.
4. **Sekul, A. A., Sparks, A. N., Beroza, M., and Bierl, B. A.,** A natural inhibitor of the corn earworm moth sex attractant, *J. Econ. Entomol.,* 68(5), 603, 1975.
5. **Traynier, R. M. M.,** Sex attraction in the Mediterranean flour moth, *Anagasta kuhniella:* location of the female by the male, *Can. Ent.,* 100, 5, 1968.

6. **Wright, R. H.,** Insect control by non-toxic means, *Science,* 144, 487, 1964.
7. **Wright, R. H.,** Some quantitative aspects of insect attraction, *Can. Ent.,* 98, 1114, 1966.
8. **Wright, R. H.,** Why mosquito repellents repel, *Sci. Am.,* 233(1), 104, 1975.

Chapter 27

SOME MEDICAL IMPLICATIONS

I presume you're mortal, and may err.

James Shirley

The human nose presents an antennal array which is as conspicuous as it is apparently unused. As a communications system its significance is only slowly beginning to receive all-around appreciation by medical men. The acetone breath of diabetes or the mercaptan breath associated with severe liver disease are an old story, to which has recently been added a special smell which some people describe as emanating from persons suffering from schizophrenia. These are obvious and positive inputs into the physicians's brain. The not-so-obvious things he will ignore at his peril, and what his nose fails to tell him — or tells him wrong — are likewise important. It is quite possible for a doctor to be influenced by a smell without realizing that his judgment has been affected. The physician's clenched fist may be a symptom of the patient's disorder.

For example, J. J. Cowley[5] exposed people of both sexes to androstenol or to a mixture of the short-chain fatty acids whose possible sexual significance was mentioned in Chapter 21. Immediately following the exposure, the people were given a psychological test which required them to form a favorable or unfavorable opinion about the suitability of various applicants for a responsible position. The results showed that women exposed to the fatty acids or the androstenol were not affected in their assessment of female applicants, but their evaluation of males was altered, especially with respect to the favorable qualities. The effect on male subjects was noticeable but less pronounced. By itself, this is no more than a danger signal.

A much more immediate hazard can arise from the fact that if the doctor cannot perceive something (especially a smell) reported by a patient, it does not necessarily mean that the patient is hallucinating. R. Asher described the case of a foundryman whose employer tried out a new chemical which the man claimed had "contaminated his blood so that the fumes were coming out all over his body."[7] It turned out that the chemical contained the element tellurium which can be absorbed into the body and then released as dimethyl telluride for weeks thereafter. The smell of this substance is one of the most offensive stinks known to man, but in the special circumstances of the man's first visit to the outpatient department, it was not noticed by the attending physician because of the way the room was ventilated. As a consequence he was almost led to a totally wrong diagnosis and came close to certifying the man as mentally impaired.

This was a case where the error arose because the patient's peculiar body odor failed to reach the physician's nose. The possibility of error would be multiplied many times if the physician is himself the unsuspecting subject of one of the abnormalities of perception such as a partial anosmia or parosmia described in Chapter 24. From the nature of his profession, he is also subject to conditions which could easily bring about other or temporary aberrations which could lead him astray. Many years ago, C. A. Elsberg[8] found that the presence of an individual in the operating room when ether was being used for general anaesthesia was sufficient to reduce olfactory acuity for several hours.

The presence of what are alleged to be olfactory hallucinations has been reported in both normal and psychotic persons. It seems clear that what would be called a parosmia in a normal person could easily be called a hallucination in one who was suspected of being mentally disturbed. If the patient says he can smell something that is imperceptible to the physician, he may be quite right. R. I. Henkin[9] found that people with cystic fibrosis were exceptionally sensitive to certain tastes and smells.

More often than not, people in whom normally subliminal sense perceptions come to the surface of consciousness are not followed up because of a strange paralysis of scientific curiosity. This paralysis is sometimes referred to as a gifted person syndrome afflicting the investigator. It is probably attributable to the lack of readily available or applicable theories and experimental techniques.

Chapter 24 also mentioned abnormalities of perception brought about by medication or by some kind of prior conditioning. One such is the aversion expressed by some people to a generalized floral scent. They associate it with funerals! To anyone unaware of the underlying reason this would be distinctly abnormal even though it was not excessively so. A more extreme but equally natural instance was that of a man who considered the smell of rotting fish to be the most pleasant of all smells. On inquiry, it turned out that his happiest times as an orphaned child were summers spent with an uncle in a New England fishing village.

As long ago as 1930, C. D. Daly, a psychoanalyst, and R. S. White, an entomologist, attempted to review human psychic reactions to olfactory stimuli and compare them with those of insects.[6] For obvious reasons, most of the emphasis was on sexual responses, but their general conclusions have been largely confirmed by later work. Male mice have been shown to respond differently to the odors of stressed and unstressed females, and it is only a few years since the discovery of the Bruce Effect, a pregnancy block in newly mated female mice if they are exposed to the smell of alien mice. This led A. S. Parkes to remark that the "attractive idea of a contraceptive perfume is no longer quite incredible."[12] Two thousand years ago, Pliny the Elder said that when the oil lamps of ancient Rome were put out the fumes "usually cause abortion".[11]

It would be interesting to enquire into the possible role of odors in relation to group phenomena such as crowd hysteria or contagious panic. A possible approach to the problem will be described in Chapter 50.

For the moment, it is sufficient to recognize that whatever the physician's olfactory perceptions may be, those of the patient ought to be treated with respect if only to the extent of trying to establish their reality. One way to do this would be through changes in the electrical conductivity of the skin — which is called the psychogalvanic skin reflex. R. W. Moncrieff suggested that even if such measurements are not a useful way to classify odors they might be a useful way to classify people.

REFERENCES

1. **Asher, R.,** Clinical sense. The use of the five senses, *B. M. J.*, 1, 985, 1960.
2. **Bruce, H. M. and Parkes, A. S.,** An olfactory block to implantation in mice, *J. Reprod. Fertil.*, 2, 195, 1961.
3. **Bruce, H. M.,** Olfactory block to pregnancy among grouped mice, *J. Reprod. Fertil.*, 6, 451, 1963.
4. **Carr, W. J., Roth, P., and Amore, M.,** Responses of male mice to odors from stressed vs. nonstressed males and females, *Psychon. Sci. Sect. Anim. Physiol. Psychol.*, 25, 275, 1971.
5. **Cowley, J. J.,** The effect of exposure to two odorous compounds (pheromones) on an assessment of people test, 1st ECRO Congr., Paris, July 3 to 5, 1974.
6. **Daly, C. D. and White, R. S.,** Psychic reactions to olfactory stimuli, *Br. J. Med. Psychol.*, 10, 70, 1930.
7. **Davidson, L. S. P.,** Mercaptan in the breath of paients with severe liver disease, *Lancet*, 2, 197, 1949.

8. **Elsberg, C. A., Brewer, E. D., and Levy, I.,** The sense of smell. IV. Concerning conditions which may temporarily alter normal olfactory acuity, *Bull. Neurol. Inst. N. Y.,* 4, 31, 1935.
9. **Henkin, R. I. and Powell, G. F.,** Increased sensitivity of taste and smell in cystic fibrosis, *Science,* 138, 1107, 1962.
10. **Henkin, R. I., Schechter, P. J., Hoye, R., and Mattern, C. T. F.,** *JAMA,* 217, 434, 1971.
11. **Montague, A.,** Those smelly roman lamps, *Science,* 163, 1271, 1969.
12. **Parkes, A. S.,** Biological control of conception, *Nature (London),* 191, 1256, 1961.
13. **Perry, T. L., Melancon, S. B., Lesk, D., and Hansen, S.,** Failure to detect trans-3-Methyl-2-hexenoic acid in the sweat of schizophrenic patients *Clin. Chim. Acta,* 30, 721, 1970.
14. **Posner, H. S., Culpan, R., and Stewart, A.,** Cause of the odor of a schizophrenic patient, *Arch. Gen. Psychiatr.,* 7, 108, 1962.
15. **Skinner, K., Smith, K., and Rich, E.,** Bacteria and the "schizophrenic odor", *Am. J. Psychiatr.,* 121, 64, 1964.
16. **Smith, K. and Sines, J. O.,** Demonstration of a peculiar odor in the sweat of schizophrenic patients, *Arch. Gen. Psychiatr.,* 2, 184, 1960.
17. **Smith, K., Thompson, G. F., and Koster, H. D.,** Sweat in schizophrenic patients: identification of the odorous substance, *Science,* 166, 398, 1969.

Chapter 28

MEDICAL IMPLICATIONS — II

In the broad field of medicine, each specialist group tends to isolate itself from other groups of specialists. When a man in one field records opinions regarding certain features of the practice in another field, it is not surprising that his opinions are likely to carry little weight and to receive skeptical consideration.

William C. Menninger

The olfactory peculiarities described in Chapter 24 can generally be associated with an ascertainable cause, such as a genetic factor, a physiological state, or the effect of medication, or diet, or smoke inhalation. Other abnormalities (or normalities, it is often hard to say which) are usually put aside as inexplicable exceptions and thereafter ignored. Given a sufficiently comprehensive view of the field, the phenomena can perhaps be understood and exploited or made the starting point for important investigations. Here we will review a few examples.

Consider whether two individuals experience the same sensation when smelling the same substance. There can be no absolute answer to this, but we do know that when a perfumer mixes different substances so as to achieve what is for him a particular fragrance, such as violet, this resemblance will be valid for most other observers. Were this not so, the manufacture of perfumes would be impossible. But there are exceptions.

The Exaltolide Phenomenon involves one of a large group of compounds having an odor very similar to that of natural musk. It sometimes happens that people who have difficulty in detecting the smell of Exaltolide can easily perceive the rather similar odor of Musk Xylol, which is chemically very different. Hitherto this sort of thing could be regarded only as a curiosity of quite unknown significance, but the way may be open to a rational explanation on the basis of overlapping vibrational patterns of the different substances.

Provided precautions are taken not to be led astray by such effects, simple threshold measurements can have some useful medical applications. More than 30 years ago, C. A. Elsberg[4] developed a special technique for measuring the minimum identifiable odor and the duration of olfactory fatigue, and showed that pathological conditions could raise or lower the threshold or alter the duration of fatigue. An analysis of 1000 patients admitted to a neurological hospital found that olfactory function was in some measure disturbed in 659 of them. Unilateral or bilateral elevation of the threshold occurred in about 40% of those with lesions in or around the frontal or temporal lobes. The fatigue studies enabled lesions to be correctly located in 54% of the patients and incorrectly in 18%. The method depended rather heavily on patient cooperation which was secured in 84% of the cases. With both threshold and fatigue measurements, localization was possible in 74% of patients with tumors in or around the frontal or temporal lobes.

Contact with the world of reality depends upon the perception of sensations and so it is important to consider how various abnormalities can react back on the person exhibiting them. Both the physician and the patient have noses and what the doctor's nose tells (or fails to tell) the doctor is only part of the story. What the patient's nose is telling the patient is a real part of his world.

The presence of what are alleged to be olfactory hallucinations in both normal and psychotic persons has often been remarked, but their possible relevance to such conditions as schizophrenia is too often ignored. Such hallucinations — if they really are hallucinations — are fairly common and not always unpleasant, the ratio of unpleasant to pleasant being no higher than 3:1. It is possible that cultural factors may underlie the fact that patients are more likely to speak of "voices" or "visions" than smells.

Hoffer and Osmond[5] believed that changes in olfactory perception are as useful in diagnosing early schizophrenia as the allegedly primary disturbances of association. They do not believe that olfactory changes should be regarded simply as hallucinations when they may in fact be the result of increased acuity to odors which are always there but not always noticed by others.

It remained for H. Wiener[10-12] to draw all these ideas into an explicit hypothesis so that its logical or empirical consequences could be explored and tested.

Starting from the premise that man emits all sorts of body odors, he assumes that some of them may transmit information to other people and can therefore qualify as External Chemical Messengers or ECMs. We are all conscious of some of these, but there are others which, for most people do not reach the level of conscious awareness, although their subliminal perception may be manifested by changes in the electrical resistance of the skin. Wiener[10-12] suggests that in some gifted persons these normally subliminal sense perceptions may come to the surface at least in part but the reality and possible significance of the fact is not taken seriously by others. "Psychiatrists and psychologists," he says, "arbitrarily accept and report the claim, or they arbitrarily conclude that the patient is neurotic and hallucinating. The idea of confirming such claims seems rarely to present itself".[10]

Assuming that a child is born with the ability, potential or actual, to perceive ECMs consciously, Wiener asks what will happen to the child as he grows up. At one extreme, nothing. He may realize that he senses things others do not, keep his mouth shut, and stay out of trouble. He may, indeed, use some of his ability to earn a living, in the cosmetic industry, perhaps, or as an unusually perceptive diagnostician or novelist. D. A. Laird[7] interviewed 254 "persons of distinction" about unrelated matters and met 8 who volunteered comments describing their ability to recognize persons and places by their odors, and in some cases their emotional response to them.

But if we assume the worst, the child born with the unusual perceptiveness may be rebuffed in all his attempts to bring his experience into consonance with others. As an infant, if he has awareness of ECMs, he uses a nonverbal language different from that of his mother, and normal bonds are not formed. Later, when he speaks he may be ridiculed when he tries to convince others of his insights, or, what is more likely, he lacks the words with which to describe his perceptions. Since these perceptions are clearly abnormal in the world he lives in they seem to him to be chaotic and meaningless, although sometimes in his encounters with others he can perceive meanings and intentions that are as correct as they are hidden from most people.

Eventually some overt conflict may bring him up against the forces of social control, most often in a mental hospital. Since his ECM output may be as abnormal as his input, it would not be surprising if a chemical test of his urine shows unusual substances. Also, a psychologist is likely to find a history of disturbed family relationships, and so on. While all this has been going on, what has been happening to the patient?

Finding his turmoil of no use, he searches for some way to explain the inexplicable, and may conclude that persecutors are responsible, or magic is at fault, or that he is different from other people, perhaps superior, perhaps even divine. In one way or another he learns to live with himself even if he cannot live with others, and he becomes withdrawn.

Summing it all up, we have a history of abnormal mother-child relationships, unstable personal relations, social hypersensitivity, and an inability to form social bonds, perceptual disturbances, and finally acute turmoil, panic, delusion formation, withdrawal, and eventual deterioration — and the name for this condition is schizophrenia.

In his conclusion, Wiener remarks, "Schizophrenia is a common, serious and mysterious disease. Its symptoms seem elusive and contradictory. I suggest that *some* schizophrenic patients possess (or are possessed by) functioning brain pathways which are absent or closed in normal men. Conscious awareness taps neurone circuits which are normally silent. If input and output of ECMs are included in this scheme, many apparently inexplicable features of schizophrenia begin to make sense. The schizophrenic patient appears to be a person whose gifts are his undoing. Properly handled, they may be his salvation."

Whether or not Wiener's hypothesis provides a correct explanation for some types of schizophrenia — and it does not claim to account for them all — one thing seems clear. Each one of us acts on the information reaching his brain and the line between what is normal and what is abnormal in smell perception is not at all plain. What is important is that experimental techniques and theoretical insights now exist that should make it possible to study systematically the physical and physiological bases of the phenomena so that, for example, what may be called hallucinations in psychotic persons may at times by indistinguishable from perfectly commonplace parosmias.

The situation postulated by Wiener is one in which the sufferer is a victim of hypersensitivity to odorous signals. At the opposite extreme, hazards may arise from the inability to receive odorous information. An all-round inability to smell things like smoke, or escaping gas, or tainted food, or other warnings is obviously dangerous and needs no comment beyond the warning (to be given more explicitly in Chapter 35) against odor control agents which operate by simply paralyzing all the olfactory sense organs. More subtle hazards could arise from the inability to receive certain specialized signals carrying information needed for the full development of the individual.

In Chapter 21 the possible existence of specialist receptors in the higher vertebrates was considered, and it was indicated that their function may be to receive information of special biological significance. Persons who lack one or more types of specialist would exhibit specific anosmias of the kind studied in detail by several observers, most notably by J. E. Amoore.[1]

It is interesting that most, if not all, of the specific anosmias so far reported relate to odorous signals that are in one way or another associated with reproductive or nutritional functions vital to the survival and continued existence of the species. It would not be surprising, then, if a person's inability to receive certain ECM might not be as much of a handicap as the hypersensitivity considered by Wiener. Accordingly, it is highly desirable that specific anosmias should be looked for in persons manifesting various psychotic conditions. A significant degree of correlation would point to a possible precondition and so open the way to appropriate preventative measures.

REFERENCES

1. **Amoore, J. E.**, Specific anosmia and the concept of primary odors, *Chem. Senses Flavor*, 2, 267, 1977.
2. **Bromberg, W. and Schilder, P.**, Olfactory imagination and olfactory hallucination, in normal and psychotic persons, *Arch. Neurol. Psychiatr.*, 32, 467, 1934.
3. **Eisenson, J., Fisichelli, V. R., and Welch, L.**, Imagined differences in the perception of identical olfactory stimuli, *J. Genet. Psychol.*, 84, 77, 1954.
4. **Ellsberg, C. A. and Spotnitz, H.**, Value of quantitative olfactory tests for the localization of supratentorial disease. Analysis of 1000 cases, *Arch. Neurol. Psychiatr.*, 48, 1942.
5. **Hoffer, A. and Osmond, H.**, Olfactory changes in schizophrenia, *Am. J. Psychiatr.*, 119, 72, 1962.

6. **Klages, W. and Klages, I.**, On the psychology and psychopathology of the sense of smell, *Arch. Psychiatr. Nervenkrankh.*, 205, 37, 1964.
7. **Laird, D. A.**, Man's individuality in odor, *J. Abnorm. Soc. Psychol.*, 29, 459, 1935.
8. **Ribert, S. L., Hollender, M. H., and Mehrhof, E. G.**, Olfactory hallucinations, *Arch. Gen. Psychiatr.*, 5, 313, 1961.
9. **Videbech, T.**, Chronic olfactory paranoid syndromes. A contribution to the psychopathology of the sense of smell, *Acta Psychiatr. Scand.*, 42, 183, 1966.
10. **Wiener, H.**, External chemical messengers. I. Emission and reception in man, *N. Y. State J. Med.*, 66, 3153, 1966.
11. **Wiener, H.**, External chemical messengers. II. Natural history of schizophrenia, *N. Y. State J. Med.*, 67, 1144, 1967.
12. **Wiener, H.**, External chemical messengers. III. Mind and body in schizophrenia, *N. Y. State J. Med.*, 67, 1287, 1967.
13. **Wright, R. H.**, Specific anosmia: a clue to the olfactory code or to something much more important?, *Chem. Senses Flavor*, 3, 235, 1978.

Chapter 29

POLICE DOGS — I

We shall never succeed in exhausting unmeasurable nature.

A. von Humboldt

Here is a typical police description of a fleeing suspect: "Male, white, 30 to 35 years, about 5 ft 8 in. tall, 140 lb, brown, shoulder-length hair, no beard, wearing blue jeans and a green ski jacket". From that, a policeman is supposed to pick him out from all the people on the street. All a police dog needs to do the same thing is to sniff some article worn or handled by the suspect for his nose to register an olfactory description that is specific to the individual and would not apply to someone else with the same haircut and wearing the same kind of clothes.

This is why nose work is the most overwhelmingly important work a correctly applied police dog does and why it overshadows such things as crowd control, harrass-and-delay, and various other manifestations of aggression. Each year, in Vancouver, Canada, (which has a uniquely effective dog squad) out of every 100 incidents where a police dog is used, at least 95 involve nose work. What is more, because of the emphasis placed on this in their training, the dogmaster-dog combination is known and respected to such an extent that whereas 90% of the normal run of offenders plead "not guilty" and go to trial, better than 90% of those apprehended with the help of a dog plead "guilty" with a substantial saving in time and expense to all parties.

Naturally, it is the dogmaster-dog combination that operates with this degree of effectiveness, and the effectiveness depends as much upon the dogmaster's understanding of the whats and hows of smells and smelling as upon the dog's ability to do the actual sniffing. Only in this way can the dog's special capabilities be applied to full advantage and the information he receives via his nose be transmitted to his master.

As part of their training, then, the dogmasters must come to understand such basic matters as the amount and kind of information carried by smell, and what happens in the dog's nose and brain when it smells things, all of which has been discussed in Part I of this book. In addition, he must know where and how the fugitive's body scent originates and how it can be conveyed to the dog's nose and how this knowledge can be exploited in such activities as tracking, area searching in the outdoors for hidden objects or persons, and for similar operations indoors.

In this and the following Chapters we will consider some of these matters because, quite apart from their significance in police work, they are as much a part of an overall view of the sense of smell as infrared spectrophotometry or controlling insect pests.

In Chapter 23 some of the problems we encounter when we try to classify smells as sensations were considered, and we came to the conclusion that it is primarily a matter of a vocabulary with some possibility of a grammar mentioned in Chapter 26.

With his brain much more highly specialized for processing smell sensations, a dog would begin by formulating the equivalent of words like "man", "dog", "bitch", "rabbit", "mouse", and so on. At a more specialized level he would formulate mental images of a particular man: "my master", or "the postman", or of a particular dog or cat or rabbit. At times, it may be important for him to register not a particular man or dog, but a particular *kind* of dog (such as a bitch in heat) or a particular *kind* of man, such as one who is angry or frightened, or who works at a particular trade or smokes a particular kind of cigarette. In other words, just as we can recognize categories of people as well as individual people, the dog can recognize categories of odors as well as individual odors.

For police purposes, there will be occasions when the individual's odor is the important thing. This is typically so when tracking a particular suspect or lost child whose personal scent has been made known to the dog.

At other times, it will be more important for the dog to recognize a certain category of odor rather than the odor of a particular object or person. In a building search, for example, he may be looking for any human scent, or, alternatively, for the strongest or most recent human scent. At other times, as when shown the scene of a crime and told to "find him!", he will be tracking a particular kind of man, namely, one in a special emotional state. The nature of this state will be discussed in the next Chapter. Meanwhile, here are actual cases to illustrate what has happened in typical incidents.

August 3, 1966. At 11:40 p.m., Vancouver Dogmaster PC 267 answered a silent alarm to an office building which had been surrounded by Patrol Units who reported hearing noises inside the building but had remained outside to avoid confusing the scent indoors. The Dogmaster entered and called out the standard warning that he would release his Dog. (This is done routinely to protect anyone who may be legitimately present in the building.) Police Dog DANNY was then released and was about to enter the furnace room when a voice called, "I'm here. I'm coming out." The burglar then emerged and was taken away after declaring that he was alone. DANNY was then ordered to continue the search. He returned to the furnace room and went to a small space behind the furnace and began to bark. A male voice called out, "Alright, call him off. I'm coming out." and the second burglar was removed. DANNY as then sent to search the yard, without result, and was taken back into the building to look for the tools used in the break-in. Once again he entered the furnace room and a third burglar surrendered! The space in which the three men were hiding was about 6-ft-long by 1-ft wide and was reached by squeezing through a narrow gap between the furnace and a hot-water heater. It could easily have been missed using man-power alone. The entire operation took 90 min.

May 15, 1971. At 4:25 a.m., Vancouver Dogmaster PC 407 with Police Dog SQUIRE was called to assist Police in a nearby municipality after a cat burglar had been scared away from a house at 4:00 a.m. On arrival at the scene at 4:45 a.m. they were shown where the intruder had been seen and SQUIRE was put on the track. After about three blocks it appeared that the track was growing weaker and the Dogmaster judged that they were backtracking along the route by which the suspect had entered the area. The Dog was returned to the starting point and picked up a new track which he followed for approximately a mile during which it passed through an area of bush, across a stream, and eventually onto a road where a suspect was being questioned by local Police. When about 500 feet downwind from the suspect, the Dog ceased to track and ran directly to him and was about to seize him when ordered "Out!" On being told the Dog had tracked him from the scene of the break-in, the suspect admitted his guilt.

January 17, 1968. At 4:00 p.m., Vancouver Dogmaster PC 387 and Police Dog SCOUT went to a warehouse where employees were holding one of two suspects. Two men had taken away a truck load of metal without arousing suspicion, but were questioned when they returned for a second. One of them escaped and disappeared toward the water-front into an extremely rough and untidy area of mud, piled rock, old timber and assorted debris. SCOUT was ordered to do an area search and eventually found the suspect concealed under a mass of rock and old timber and quite invisible from above. He was charged with "Theft Under $50.00" whereupon he stated, "Those sure are good Dogs. They should have more of them!"*

* From City of Vancouver Police Department Dog Squad, Various Reports, With Permission.

REFERENCES

1. **Kalmus, H.,** The discrimination by the nose of the dog of individual human odors and in particular of the odors of twins, *Br. J. Anim. Behav.,* 3, 25, 1955.
2. **Kloek, J.,** The smell of some steroid sex hormones and their metabolites, *Psychiatr. Neurol. Neurochir.,* 64, 309, 1961.
3. **Nicolaides, N.,** Skin lipids: their biochemical uniqueness, *Science,* 186, 19, 1974.
4. **Romanes, G. J.** Experiments on the sense of smell in dogs, *Nature (London),* p. 273, July 21, 1887.
5. City of Vancouver Police Department Dog Squad, various reports.

Chapter 30

POLICE DOGS — II

He who learns the law of phenomena gains not only learning, but the power of entering into the course of nature and of working on it further according to his will and need. He gains insight into the future course of these phenomena. He gains, indeed, faculties that in superstitious times were looked for in prophets and magicians.

A. von Helmholtz

September 2, 1971. 4:00 a.m. During a refresher-training exercise, a PC laid a trail along a city street and through a lane adjoining a junk-yard. Vancouver PC 248 with Police Dog SHEAFER then commenced a track but half way along, the Dog became excited and diverted into a hole in the fence surrounding the junk yard and searched without result. The proprietor was called and items worth $1200.00 were missed and found in an adjoining yard where they had apparently been abandoned on the approach of the Police. This is an example of an experienced Dog recognizing and responding to the special body emanation associated with an act of wrong-doing.*

Given that experienced police dogs which have apprehended a few real offenders have learned to recognize a special emanation from persons in that category, it is necessary to enquire what scientific basis presently exist to explain it. In short, can a frightened-hostile-angry-resentful person emit a special array of chemical molecules which the dog's nose can recognize as something his master considers important?

At the moment, we cannot give a list of the chemical substances which together give rise to what may be called the "fugitive-suspect" smell, but by putting together data from a variety of scientific disciplines we can say that the dog's responses can be explained without calling upon any unknown, parascientific, instinctive, or other essentially mysterious mechanism.

In his book, *Human Perspiration*, Professor Yas Kuno[7] describes several kinds of sweating. First there is insensible perspiration which is independent of specialized sweat glands and results from simple diffusion of water molecules and other substances, such as carbon dioxide, outward through the skin. It may therefore be expected to include other chemicals present in the blood and to respond fairly quickly to changes in its composition, more especially if there is an increase in the peripheral circulation, as when someone blushes. One such chemical change is the almost instantaneous release of adrenalin into the bloodstream when a person is frightened or furious. To the human nose, adrenalin is usually odorless, but it or substances which accompany it may well be perceptible to the more discriminating and sensitive canine nose and so may form *part* of the special smell pattern we are discussing.

In addition to the invisible and insensible perspiration, there are two types of sensible perspiration. Sensible perspiration originates in sweat glands of various kinds, located in various parts of the body, and activated by different messages emanating from the controlling Central Nervous System (CNS).

The most familiar kind of sensible perspiration is thermal sweating which takes place in warm surroundings or when one is working or exercising strenuously. It does not usually start instantaneously but only after a latent period of a minute or two, and it takes place over the whole body surface *except* the palms of the hands and the soles of the feet. The second kind of sensible perspiration is called mental sweating. It is not associated with a heated environment or in any direct way with exercise, *and it has no latent period.* It is a practically instantaneous response to almost any sort of

* From City of Vancouver Police Department Dog Squad Report. With permission.

mental stress. It takes place on the palms of the hands and the soles of the feet (which do not exhibit thermal sweating) and also appears on the face and in the armpits but not, as a rule, elsewhere on the body. It is most readily detected by a change in the electrical resistance of the skin, the so-called "psychogalvanic skin response" which is one of the responses recorded by the polygraph or lie-detector.

There are many kinds of mental or emotional stress which can produce mental sweating. For experimental purposes the most painless and easily applied stress is an exercise in mental arithmetic. This will usually induce sweating on the hands and feet but not usually on the forehead. Other kinds of stress range from reading a book to eating hot chili or being pricked with a pin or receiving unexpected or unwelcome news.

The facts that (1) mental sweating has no latent period and starts within 1 or 2 sec. of the application of the stress and (2) the rate of secretion depends upon the intensity of the stress and persists as long as the stress continues, together show that mental sweating is stimulated by a part of the CNS different from that which regulates thermal sweating. The biological function of this sort of response has remained almost wholly unexplored and unknown, at least at the human level. At the level of insects, ants and bees regularly transmit information throughout the nest by chemical signals, and various animals such as rats, mice, rabbits, and dogs control at least some of their social activities by means of chemical messengers. Therefore it is not unreasonable to suppose that at the human level we retain some vestiges of a primitive ability to pass information from person to person by means of body emanations. Moncrieff,[10] for example, showed that some people exhibit a psychogalvanic skin response on presentation of an odor.

For reasons partly cultural and partly historical, the role of External Chemical Messengers (ECMs) in relation to human behavior has receded into the background. For one thing, we walk on our hind legs and our noses are several feet removed from the ground. For another, as soon as men recognized the nature of seeds and established settled communities based on agriculture, the communities emitted a foul odor. When large groups occupy small areas and have not developed the facilities needed for personal and communal sanitation, the result has to be a downgrading of the olfactory sense insofar as its ability to receive subtle hints and perceive delicate nuances is concerned. Only within the past couple of centuries has urban technology begun to correct this situation: the first indication being the appearance of very graceful and refined French minuets, *danced at arm's length!*

For cultural reasons, any mention of personal body odors has become a forbidden subject in polite society, but there are individuals who have claimed the ability to know the feelings and emotional states of others by using their noses. A psychiatrist, for example, has described the odor of fear as, for him, "a rather strong, bitter smell", and an odor associated with general anxiety as being, "sour, stale, rather vinegary". A professional perfumer who was asked whether he could distinguish the odors of various moods such as fear or anger replied that of course he could, and seemed surprised that the question was worth asking.

From all this, it seems clear that an experienced police dog may well become able to recognize that certain individuals are in a certain emotional state on the basis of their body odor. What remains to be established is the identity of the emanations which convey the information. With the growing refinement of techniques of gas-liquid chromatography this is no longer a mere vague possibility.

REFERENCES

1. **Brown, W. L., Eisner, T., and Whittaker, R. H.,** Allomones and Kairomones: trans-specific chemical messengers, *Biosciences,* 20, 21, 1970.
2. **Bruce, H. M.,** Effects of olfactory stimuli on reproduction in mammals, in Ciba Foundation Study Group No. 26, 1967.
3. **Comfort, A.,** Likelihood of Human Pheromones, *Nature (London),* 230, 432, 1971.
4. **Curtis, R. F., Ballantine, J. A., Keverne, E. B., Bonsall, R. W., and Michael, R. P.,** Identification of primate sexual pheromones and the properties of synthetic attractants, *Nature (London),* 232, 396, 1971.
5. **Frosecchi, M. and Digiesi, V.,** Influence of odorous stimuli on peripheral circulation in man, *Rass. Neurol. Veg.,* 18, 100, 1964.
6. **Karlson, P. and Luscher, M.,** Pheromones: a new term for a class of biologically active substances, *Nature (London),* 183, 55, 1959.
7. **Kuno, Y.,** *Human Perspiration,* Charles C Thomas, Springfield, Ill., 1956.
8. **Laird, D. A.,** Man's individuality in odor, *J. Abnorm. Soc. Psychol.,* 29, 459, 1935.
9. **Michael, R. P., Zumpe, D., Keverne, E. K., and Bonsall, R. W.,** Neuroendocrine factors in the control of primate behavior, *Rec. Prog. Horm. Res.* 28, 665, 1972.
10. **Moncrieff, R. W.,** Psycho-galvanic reflexes to odors, *Perf. Essent. Oil Rec.,* 54, 313, 1963.
11. **Schaffer, J.,** Odor producing organs of the human body, *Wien. Klin. Wochenschr.,* 50, 790, 1937.
12. **Weiner, H.,** External chemical messengers. I. Emission and reception in man, *N. Y. State J. Med.,* 66, 3153, 1966.

Chapter 31

SCENT TRACES (POLICE DOGS — III)

I do not know whether detective stories are permissible evidence in a scientific discussion; but I read them and I wish to close with a quotation from one. "The evidential value of any fact is an unknown quantity until the fact has been examined."

W. D. Bancroft

Body odor emanating from the skin surface can be transferred to other surfaces by contact, when an object is handled or accidentally brushed against or stepped upon. This enables a dog to pick out objects handled by a particular person after smelling that person's hands or, alternatively, to recognize the owner of a particular object. The amount of material transferred from the body to the object can be almost unbelievably small but, thanks to the amazing sensitivity of the nose, the scent left behind by even the most careful handling can still be perceived after a surprisingly long time.

Experimenters at Duke University prepared trays carrying four carefully cleaned glass microscope slides together with a fifth slide which had been given the imprint of a single finger. Some of the trays were stored indoors. At intervals ranging from 6 hr to 3 weeks, trays were removed and tested to see whether a dog could pick out the slide with the fingerprint. One hundred tests with a fox terrier showed 94 correct choices after 3 weeks. A similar test with the slides placed on the laboratory roof and exposed to blown dirt, direct sunlight and rain gave 77% correct selections after 24 hr and a still-significant 45% after 1 week, using the same fox terrier.

A footprint on the ground includes things like leather, disturbed earth, crushed grass, and substances previously trodden upon, as well as the unique personal contribution from the owner of the boots. The amount of this personal contribution has been estimated in an interesting calculation by Professor Neuhaus[9] in Germany.

The sole of the foot has only sweat glands, but in substantial number: about 600/ cm^2. Hence, for the overall smell of the feet, sweat glands are the important contributors. Over 24 hr the human body may secrete 800 cm^3 of sweat. From its 2 million sweat glands the sole of each foot would release about 2% of this daily production, or about 16 cm^3. Also, human sweat has approximately 0.04% of fatty acids of the type of valeric acid (old running shoes smell). It can be assumed that 1/1000 of the fatty acid penetrates steadily through the sole and seams of the shoe and that the foot rests on the ground for half a second at each step, and that the total fatty acid fraction is made up of ten different fatty acids. These are minimum assumptions. From this, one can calculate that of any one fatty acid, such as valeric acid, about 4×10^{-11} g will be deposited and left behind in each footprint. This is 50 to 100 times less than the amount in a single fingerprint, but there is still enough to account for the ability of a properly trained dog to recognize and follow the track of a single individual even when it has been entangled with the tracks of others.

G. J. Romanes[13] once lined up 12 men one behind the other with himself in the lead and they set off with each man treading carefully in the footprints of the man in front. After going about 200 yards, the party split with Romanes and five men going off to the right and the other six to the left. After they had gone a considerable distance they all hid and Romanes' dog was put on the trail and told to find her master. She did so with only a slight check when she overshot the place where the party divided.

This sort of thing is made possible by two things: the sensitivity of the nose and the ability to discriminate between closely similar odor patterns, both of which have been studied in an elegant series of experiments by Professor Neuhaus.[6-11]

Table 1
OLFACTORY THRESHOLDS: DOG VS. MAN

Compound	Molecules per 50 cm^3 sniff — Dog	Man	Ratio Man/dog
Butyric acid	4.5×10^5	5.0×10^{11}	1,100,000
Valeric acid	1.8×10^8	7.3×10^{11}	410,000
Ethyl mercaptan	1.0×10^7	6.3×10^{11}	63,000
Ionone	5.0×10^6	1.2×10^{10}	2,500

Note: Human thresholds from Laffort, Dog thresholds from Neuhaus. A recent comparison of the Ionone Thresholds for six humans and three dogs gave a ratio of about 300 or about ten times lower than the value calculated above.[5]

He contrived an apparatus and technique for measuring very precisely the olfactory threshold concentrations of various substances that are just perceptible to a dog. For most odorous substances, the dog is more sensitive than a man by a factor of the order of 1000, but for the fatty acids which contribute so much to the odor of perspiration, the dog can be as much as a million times more sensitive, as shown in Table 1. There, the concentrations are expressed, perhaps somewhat unconventionally, in terms of molecules per sniff, one sniff being put equal to 50 cm^3, or the approximate volume of air drawn in during a single sharp inhalation or sniff. Returning to Professor Neuhaus' calculation of the amount of valeric acid left behind in a footprint, namely, 4×10^{-11} g, this weight contains no fewer than 2.5×10^{11} molecules which would be sufficient to be perceptible to a dog when dispersed in 140,000 50 cm^3 sniffs. Indeed, the table shows that a somewhat better-than-average human nose should be just able to detect a footprint when it is fresh, and Professor Neuhaus states that he could do this himself when a person walked once across a piece of clean filter-paper laid on a hardwood floor.

In other experiments using mixtures of four fatty acids, Professor Neuhaus showed that the addition of a fifth in an amount too small to be perceived by itself, nevertheless altered the smell of the mixture enough for a dog to discriminate the mixture carrying the added trace. An odor as complex as that of a footprint on the ground can be altered by subthreshold amounts and assorted traces of chemical molecules which together determine both the total olfactory input pattern, and the subpatterns or by-odors the nose-plus-brain can single out for attention.

Before a dog can be trained to the level of performance required for police or other professional levels of competence, it must be put through a thorough and appropriate educational process. Otherwise, he will take the easy course of following a visible set of footprints and the more easily perceived smell of disturbed earth and crushed grass. Thus, when Colonel Most[4] of the Berlin police had a man walk halfway across a field of rather soft earth the track was made to disappear into thin air by carrying the man away with a trolley attached to an overhead cable. The track was, however, continued onwards by a large wheel with shoes fixed at one-pace intervals around the rim. An untrained dog set to follow the trail continued past the changeover point without apparently noticing any difference.

An ever-present pitfall in all experiments of this sort is called the "Clever Hans Error". Clever Hans was a horse whose owner believed him to be able to do sums in simple arithmetic in his head and tap out the answer with his hoof. On investigation, it turned out that the horse could tap out the answer correctly only if it could see someone who knew both the question and the answer and who could, quite without meaning to, signal the horse to start and stop tapping. The most usual signal was a slight bending forward to start the horse tapping and an equally slight straightening up when the correct number of taps was reached.

In the same way, one dogmaster-in-training would put his hands on his hips when he thought his dog was searching in the wrong direction, and another would rub his chin, and both dogs soon learned to recognize and respond to these visual directives and allow them to override what should have been their primary duty to use their own noses. In at least one case, a dog was so sensitive to his master's wishes that orders could not be given verbally without the tone of the voice conveying unintended messages; the solution to the problem was to train the dog to respond to long and short blasts on a whistle.

Not every dog can be trained to these levels of performance and not every man or woman is temperamentally capable of doing the training. But given the right dog and the right dogmaster, and with proper precautions against the Clever Hans type of error, the olfactory capabilities of the vertebrate nose can be explored, demonstrated, and used.

The fact that when we look for the chocolate contribution to a combination of chocolate and peppermint, we usually take a fresh sniff at the sample, may mean that our Central Nervous System (CNS) can set up a sort of filter to let only part of the signal make its way up through the network of nerves during the sniff. Figure 2, Chapter 22 shows nerves coming down from the higher levels to the lower ones as well as the nerves leading the other way, and this may be their function. We do not yet know the details, but we do know this: the ability of the brain to do any particular job can be improved by practice and training, and what is true for us is just as true for the dog.

His ability to register, analyze, and respond selectively to the various components and contributors to a complex smell can be developed and improved by training. The machinery is there, but it will never function if its bearings never become run in and if its joints are rusted because they have never been flexed.

REFERENCES

1. **Campbell, P., Laughy, P., and Wright, R. H.,** *Family Dog Training for Fun, Security and Profit,* Douglas, David and Charles, Vancouver, 1975.
2. **King, J. E., Becker, R. F., and Markee, J. E.,** Studies on olfactory discrimination in dogs. III. Ability to detect human odor trace, *Anim. Behav.,* 12, 311, 1964.
3. **Laffort, P.,** Attempted standardization of the human olfactory thresholds for 192 pure compounds, *Arch. Sci. Physiol.,* 17, 75, 1963.
4. **Most, K.,** Experiments with human-like but non-human odor trails, *Hund.,* 469, 505, 1927.
5. **Moulton, D. G. and Marshall, D. A.,** Odor detection in dog and man, 1st ECRO Congr., Paris, July 3 to 5, 1974.
6. **Neuhaus, W.,** The odor thresholds of dogs for fatty acids, *Z. Vgl. Physiol.,* 35, 527, 1953.
7. **Neuhaus, W.,** The discrimination of odor quantities by men and dogs in experiments with butyric acid, *Z. Vgl. Physiol.,* 37, 234, 1955.
8. **Neuhaus, W.,** The odor threshold of odor-mixtures for the dog and their relation to the thresholds for unmixed odors, *Z. Vgl. Physiol.,* 38, 238, 1956.
9. **Neuhaus, W.,** The ability of dogs to discriminate mixtures of odors, *Z. Vgl. Physiol.,* 39, 25, 1956.
10. **Neuhaus, W.,** The thresholds of perception and recognition in the olfaction of the dog in comparison with the olfactory perception in man, *Z. Vgl. Physiol.,* 39, 624, 1957.

11. **Neuhaus, V.**, Differences in the acuteness of olfaction in dogs, *Z. Vgl. Physiol.*, 40, 65, 1957.
12. **Pfungst, O.**, *Clever Hans (The Horse of Mr. von Osten)*, Holt, Reinhart & Winston, New York, 1965.
13. **Romanes, G. J.**, Experiments on the sense of smell in dogs, *Nature (London)*, 36, 273, 1887; see also **Linn, J.**, *Soc. (Zool.)*, 20, 65, 1890.

Chapter 32

TRACKING (POLICE DOGS — IV)

Learning, undigested by thought, is labor lost; Thought, unassisted by learning, is perilous.

Confucius

The track left on the ground by a person walking over it may be visible as well as smellable. The smellable part includes a personal scent and also the smell of disturbed earth, crushed grass, and broken plants. With the passage of time the personal contribution fades faster than the rest and the way a trained tracking dog performs will change accordingly. In a rural setting, or in a forest, the track of the quarry is unlikely to be criss-crossed by other tracks, and so a dog can still do a reliable track after a surprisingly long time. The coldest trail known to have been followed correctly by a dog belonging to the Royal Canadian Mounted Police was 56 hr old and almost certainly could have retained little or nothing of the personal scent of the quarry.

In a city or built-up area, the conditions are different in two distinct ways. The track is almost certainly crossed by many others, but the dog usually arrives on the scene within an hour and seldom more than 2 or 3 hr after the quarry has departed. He has, therefore, a much better prospect of picking out one track from all the rest, and experience has many times demonstrated that, given proper training, it can be done. Perhaps the most striking demonstration of this was an experiment in which a Vancouver police dog tracked a particular person across a playground during the half-time interval in a game of soccer football.

The details of the process have been studied in a very careful and systematic way by H. M. Budgett,[1] a Master of Foxhounds in England. His results are useful both for the scientific way he made his experiments and because he was mainly concerned with rather fresh tracks where the personal scent is significant.

Several kinds of evidence indicate that a track is formed only when the quarry makes actual physical contact with the ground or the vegetation, or when he touches or brushes against such things as walls, gates, or bushes, or, what would be equivalent, if he sheds skin cells (sometimes called "rafts") onto the ground from his body surface. It is clear, however, that very few rafts would be shed by a fully clothed man wearing boots and long underwear, though they could contribute substantially to the track of an athlete in running shorts or of a bare-legged girl wearing sandals.

Volatile odorous vapors released from matter deposited in these ways form an invisible scent-cloud in the vicinity of the deposited material for a time that may be longer or shorter depending upon the wind, weather, and the nature of the ground. At times, the cloud will be breast-high and the dog will run with his head up; at others he may have to keep his nose close to the ground and seek out each footprint. It is a useful exercise for a dogmaster to have someone put a strong-smelling substance, such as oil of wintergreen, on his boots and for the dogmaster to cast dignity aside and try to track his quarry on all fours. The exercise should remove the magic and at the same time increase his respect for his dog's capabilities.

The effect of weather conditions on the scent track can be summed up in a few rough rules.

1. If there is a wind blowing across the actual track, the scent cloud will be on the downwind side of the track and the dog will run parallel to rather than on the track. Naturally, if the wind is very strong the track will have to be very strong and fresh if it is to be followed. Sometimes, if the track runs along a road with

a ditch on the downwind side, there will be a relatively still eddy in the ditch and the dog's behavior may suggest that the quarry had actually moved along the ditch rather than on the road.

2. If the wind is blowing in the same direction as the track, the dog is likely to follow it even with a longer start and a fairly brisk wind, unless the ground is so uneven or interrupted by trees and buildings as to create much turbulence and mixing. This is especially true if the dog is tracking in an upwind direction so that the scent is brought directly to his nose from many footprints. For the same reason, if a dog comes upon a track from the side he will usually turn and track in the upwind direction if given the choice. Since the scent track itself has no directional arrow to show whether the quarry was moving with the wind or against it, the dogmaster may have to turn the dog around if visible footprints or other information tells him the track should be followed downwind.
3. Apart from air movements due to the wind, the strength of the track will vary with the nature of the ground. On hard ground — pavements or well-trampled playgrounds — the scent is likely to be weak, and if the surface is dusty the dust may make it hard for the dog to use his nose at all. On soft earth the scent is likely to be better because deeper footprints mean a larger area of contact between the quarry and the ground. This is even more so with grass or shrubbery, but with the possible complication that the dog may pay more attention to the bruised-earth smell than to the specific scent of the quarry. To test this during training exercises, the quarry can lay part of the track using his own shoes and at some point change them for clean rubbers or, perhaps, for shoes belonging to someone else.
4. On wet ground the scent is likely to be strong for two reasons. First, wet ground is usually softer and the footprints deeper; and second, oily components of the scent will tend to spread over the watery surface and expose a larger surface area for evaporation. Naturally, if it is raining so heavily that the water is visibly running off the surface the scent will soon be carried away entirely.
5. Temperature conditions can affect the track in two ways, namely, the condition of the track at any moment, and the lifetime of the track after it has been laid. Tracking under city conditions is likely to involve rather short starts, so that track persistence is not as important as track strength. For any given kind of surface, the scent will generally be stronger when the ground is warmer than the air immediately above it, and the scent will be weaker and the track harder to follow when the ground is cooler than the air. Using two thermometers, one in the ground and the other a few inches above it, H. M. Budgett showed that hounds tracked best when the temperatures were two or three degrees "to the good" (ground warmer than air); they tracked fairly well when the temperatures were about the same, and rather poorly when the temperatures were a few degrees "to the bad". As he put it, "the earth exhales when its temperature is higher than that of the air, and inhales when it is cooler than the air".[1]

Without actually measuring the air and ground temperatures, it is often possible to estimate the chances of a successful track by remembering a few general principles.

1. At sundown the air generally cools faster than the ground so that during the night the scent-tracking is likely to be good. The temperature an inch below the surface is not likely to change by more than a degree or two in several hours, whereas the air temperature may change by as much as 15°.
2. Fog usually results when cold air moves over warm, moist ground, so that scent-

ing is likely to be very good in a fog unless it is heavily polluted with engine exhaust and other fumes.

3. Where the ground is sheltered by bushes or buildings, the scent may be good or bad depending upon what has happened in the hours preceding. For example, the scent may be good in the open but bad under shelter if the ground in the open has been warmed by the sun and the air over it chilled by a cool breeze. Meanwhile, in the sheltered place, the temperatures will be more nearly equal where the sunlight has not penetrated.
4. The nature of the ground can affect the rate at which its temperature can rise or fall. Paved surfaces will accommodate their temperatures faster than porous or grassy ones. Wind will accelerate heat transfer by convection, and the humidity of the air has an important bearing on the rate of heat loss by the ground because water vapor absorbs infrared radiation very strongly and dry air hardly at all. After a warm day, scenting is likely to be better in the later hours of a cloudy night than a clear one.
5. Strong sunlight will shorten the life of the track both by hastening evaporation and by chemically altering the chemicals left behind by the quarry. During daylight the individual's personal scent probably does not last more than 5 hr in most cases. Exceptionally, it may last longer, but it is difficult to be sure because of the contributions from such partly individual contributors to the scent as shoe polish.
6. Quite apart from the state of the weather and the nature of the ground, the strength of the track can depend upon the way the quarry moves. One who shuffles or drags his feet, or one who takes short steps or carries a heavy weight, or brushes against bushes or walls, will leave a stronger track than one who moves lightly and lifts his feet smartly. Finally, the state of mind of the person making the track can modify the nature of the track and conceivably its strength as well, for reasons explained in Chapter 30.

In Chapter 1, osphresiology was described as a science for all seasons. This is never truer than when we take a really comprehensive look at the intricacies of tracking by scent.

REFERENCES

1. Budgett, H. M., *Hunting by Scent,* Eyre and Spottiswood, London, 1933.
2. Syrotuck, W. G., *Scent and the Scenting Dog,* Arner Publications, New York, 1972.

Chapter 33

SEARCHING (POLICE DOGS — V)

I sometimes search the grassy knolls for wheels of hansom-cabs.

Lewis Carroll

Attempt the end, and never stand to doubt; Nothing's so hard, but search will find it out.

Robert Herrick

Searching differs from tracking in that the person or object being sought stays in one place but releases a trail of scent which is wafted away in currents of air. A properly trained dog will cruise to-and-fro until he picks up the scent and will then try to follow it back to its source. Estimates of the downwind distance at which the scent can be detected vary from a few yards to as much as a mile. As already mentioned in Chapter 31, a single fingerprint constitutes a significant source of scent so that an object which has been worn or handled will be a fairly strong source, and a living, breathing, and possibly agitated person will generate a really strong scent.

Whatever the source, the scent will be carried away by moving air currents and will be progressively diluted and dissipated in the process. Contrary to what might be thought at first, a strong wind will not carry the scent as far as a gentle one for the following reason: a 10 mi/hr wind moves ten times as much air past the source as a 1 mi/hr wind, and so dilutes the scent ten times as much right at the source. However, while the 1 mi/hr wind will carry the scent farther, it will take a full hour to move it a mile, during which the wind must blow steadily and without changing its direction. Similarly, in an indoor search, an air movement of 1 cm/sec will move the scent no more than 2 ft/min, and correspondingly longer to carry it across a room, along a corridor, or from one floor to another. Thus, the longer the time which elapses before a search begins the better the chance of success.

It is sometimes supposed that odors stay close to the ground because the odorous molecules are heavy. The fact is that when a small amount of odorous material is added to the air, the effect on the overall density is quite insignificant. For example, camphor can just be smelt at a concentration of 1.13×10^{-7} g/ℓ of air. At a concentration 1000 times higher, the amount of camphor in a liter of air is 0.0001 g. Since a liter of air at ordinary conditions weights approximately 1 g, the overall change in the density is only one part in 10,000. Therefore, odorous molecules are carried passively wherever the air takes them.

In the open this usually means the wind, but it can be modified by the effects of convection if the ground is unevenly heated by the sun. For example, I have seen the smoke from a factory situated at the foot of a hill next to a lake, being carried up the hillside as long as the sun was shining on it, and, within a few minutes, reverse its direction and spread over the lake when a cloud cast its shadow on the hillside. On a smaller scale the same sort of thing can happen near the shadow of a tree or building. The most difficult condition for an open air search is when the quarry is hidden at the top of a bare, rocky knoll which is heated by the sun so that his scent is carried up and away.

Qualitatively similar considerations apply when searches are conducted indoors, but with a great quantitative difference in the usual rates of air movement. Indoors, and with no forced circulation of air, the currents are seldom stronger than a few centimeters per second, because they are are a result of simple convection. Warm air is lighter than cool and so tends to float upward. Similarly, moist air is lighter than dry air because the molecular weight of H_2O is only 18 as compared with 28 for N_2 and 32 for O_2. An increase in relative humidity from 50 to 60% has the same effect on the density as a change in temperature of 0.25°C and is enough to create a slight updraft.

On a warm, sunny day after a cooler night the temperature inside a building is likely to be cooler than outdoors, and so air will enter the building near the roof and sink down and finally make its way out through the basement area. At such times a search should begin downstairs and work its way up. On a cold day in a heated building, or at night after a warm day, the air movements inside will be reversed.

These general principles apply to the building as a whole. Within a single room there will be a local circulation so that, for example, on a summer evening after a warm day, the interior walls and objects inside a room will retain some warmth after the windows and outer walls begin to cool. Thus there will be an updraft to the ceiling, across the ceiling to the outer wall, down past the windows, and back across the floor. Under these conditions, a dog may indicate the scent near the windows when the quarry is actually hidden in or on top of a cupboard and a long way from where the dog registers the scent. If it is a warm day outside, or if there are radiators under the windows, the movement will be just the opposite.

Besides circulation inside the building, there is usually a quite surprising amount of interchange of air between indoors and outdoors even when the windows and doors are all closed. Before World War II, scientists at Cambridge University in England made some experiments to test the probable effectiveness of the government's instructions for making a gas proof room in a private house. To do this, they shut themselves in and then raised the level of carbon dioxide in the room by smashing up a block of dry ice, and then measured the concentration of CO_2 at intervals thereafter. Generally speaking, in an unsealed room half the gas had leaked out and been replaced by air from outside in less than an hour, and if it was windy outside, the times became very short. The tests were all done with the experimenters inside and the lights on, so that there was more air movement than there would have been in an unoccupied house.

What these experiments show is that it is quite possible for a dog to know that there is someone concealed in a building even though the doors and windows are shut, provided there has been enough time for the scent to be carried to the outside by air currents. A dog at ground level will be most likely to do this when the air inside the building is cool so that the circulation is in at the top and out at the bottom. Similarly, in one of their experiments, the Cambridge group had a man carry a piece of cloth soaked in strongly smelling banana oil past the outside of a house during the evening, and found that those inside would perceive the smell within a couple of minutes if they were close to the sealed-up windows.

It is usually very difficult for a human observer to detect the speed and direction of air movements resulting from simple convection unless he uses smoke or some other means to make the movement visible. Thus it is usually hard for him to track a smell to its precise source using his nose alone.

A dog can somehow avoid this difficulty. For example, he will indicate upwards when the scent is moving downwards beside a cold wall, or locate the source of a scent concealed in a box or under a rug. Evidently, he has some means of registering these extremely gentle air movements so as to know from which direction the scent is coming. At the time of writing there do not seem to have been any studies to identify the sensory organs which give him this information, but the most likely source would be the long flexible whiskers on each side of his face and nose. They would tend to bend more equally when he is pointing straight upwind and would be sensitive to even the weakest air currents.

Chapter 34

AIR POLLUTION

The air is a confused aggregate of effluviums from such differing bodies that, though they all agree in constituting by their minuteness and various motions one great mass of matter, yet perhaps there is scarcely a more heterogeneous body in the world.

Robert Boyle

The atmosphere can be polluted in various ways: with dust or liquid droplets, with corrosive or irritating chemical vapors, or simply with bad smells. It is usually these last which provoke the loudest complaints and are the most difficult to treat by legal enactments. The difficulty has been nicely summed up by Sir James Fitzjames Stephen:

> . . . another remarkable matter in connection with the law of nuisance is the nature and extent of the limitations upon it. It is said in an old case in regard to candle-making in a town, "Le utility del chose excusera le noisesomeness del stink"; and a law which required in a large town the quietness and purity of air which may fairly be expected in the country would be absurd, and inconsistent with the common interest. This principle is unquestionable law, but it requires great care and consideration in its application, for it is limited by others which are not obviously and perhaps not really consistent with it on all occasions.*

Industries are concerned with the atmospheric environment at three levels of significance: as a matter of good public relations, or in order to conform with laws and regulations limiting certain emissions, or because they may be sued by third parties.

Thus, in addition to proven or suspected health hazards associated with this or that pollutant, there may be a largely subjective annoyance which is not demonstrably pathogenic but which does interfere with an individual's comfort and well-being. At the very least it is a matter of public relations interest and may become more than that if attitudes in the community redraw the line between health and disease. In Sweden, annoyance is considered a medico-hygienic problem and some studies have been made to find out how badly people are annoyed by factory smells. In one survey, about 20% of people living near a factory making wood pulp reported major annoyance, and it was perhaps significant that many of them had previous respiratory or cardiovascular disease, or had showed a propensity to neurosis and a sensitivity to airplane noise, and so on.

Nevertheless, it is clear that as a public relations problem, bad smells (which are essentially smells that are unwanted at a particular time and place) constitute an annoyance wherever and whenever they are perceptible. As a consequence, the objective of control measures is to minimize the afflicted area while recognizing that within the area there will always be some degree of nuisance. As a practical measure, standards are set to restrict the downwind distance at which the nuisance will be felt, but because of the unpredictable nature of moment-to-moment air movements it is never possible to assert that a particular emission standard will infallibly protect a given locality. During a period of calm, even a small release of smelly material can build up a localized cloud of smoke or smell which can later move off and be carried a very long way by a gentle breeze.

Because the nose is so sensitive, chemical tests will often fail to detect things that are instantly apparent to a normal person. For monitoring odorous air pollutants, human observers are therefore indispensible. In a Swedish study, ambient odors were

* From Stevens, Sir James Fitzjames, *A General View of the Criminal Law of England,* London, 1890, 106. With permission.

Table 1
DOWNWIND TRAVEL OF AN INDUSTRIAL ODOR

	Distance (km)			
	(2)	(5)	(10)	(20)
Observed odor frequency, %	10.8	9.8	8.5	5.1
Predicted odor frequency, %	9.1	5.7	3.2	1.7
Number of observations[a]	6428	7490	5528	6972

[a] Momentary judgments every 2 min.

recorded by persons either moving about in automobiles or stationed downwind from a pulp mill. The results revealed a complex exposure pattern, and what was termed a surprising lack of homogeneity in the gas plume, probably attributable to clusters or filaments of odorants. What had previously been thought to be incredible discrepancies in the pattern of complaints turned out to be fully explicable. Most important, the problem existed at much greater distances than could be predicted by the usual methods of mathematically estimating atmospheric dispersion and mixing, as shown in Table 1.

It is clear that in public relations terms, industry faces a continuing problem. At one time there was a tendency to deny that any problem existed, or to blame it on someone else, or, ultimately to justify it with comments such as, ''It smells like money!'' More recently, at least some industries have begun to take the public into their confidence by admitting the existence of a problem and making it clear that genuine efforts are in train to control it. A Mexican firm even went to the length of bringing housewives from the surrounding area into the factory so that they could learn to recognize the specific malodors from each part of the operation and so help the operators to concentrate their efforts on the most significant emissions. It would be difficult to imagine a more impressive demonstration of management's good faith and genuine desire to improve the quality of life. It is, unfortunately, an almost unique instance. It is much more usual for the community, as an organized entity, to take the initiative.

The community approach to pollution control may be aimed directly at the perceived nuisance, or it may simply take the form of laws and regulations designed to keep the release of certain specified substances below specified limits. The Offensive Odor Control Law in Japan is an example of the first approach. It designates ammonia, methyl mercaptan, hydrogen sulfide, and several other substances as offensive odorants and classifies odor intensities in six grades from zero (odorless) to grade 5 (the most offensive smell) and sets limits to the odor level permissible in industrial and residential areas.

Where the approach is to specify allowable emissions, two considerations operate. The first seeks simply to restrict the affected area, and the second seeks to ensure that substances with potentially harmful effects are not released in dangerous amounts. The estimation of potential hazard is relatively easy with gaseous emissions of known composition because toxic concentrations can be measured with exposure tests with plants or animals. However, even though polluting substances may not be harmful when taken singly, when presented in combination the effect may be more severe. Thus in a U.S. study,[6] rats, mice, and monkeys were exposed during 90 days to ''Industrial Threshold Limit Values'' of hydrogen sulfide, methyl mercaptan, indole, and skatole. Mortality rates were high only in the group exposed to a mixture of the four compounds.

Table 2
INCIDENCE OF CERTAIN PATHOLOGICAL CONDITIONS IN RELATION TO AIR POLLUTION

	High pollution area		Low pollution area	
Pathology	1970 (%)	1971 (%)	1970 (%)	1971 (%)
Chronic paranasal sinus	9.9	13.5	3.8	7.9
Chronic rhinitis	10.6	8.2	10.3	4.5
Nasal allergy	5.2	4.9	2.7	2.4

Besides the possible effects of the simultaneous presentation of two or more substances, one must not overlook the possibility of new compounds being formed under the influence of sunlight or by reactions with emissions from unrelated sources. Air pollution in the Los Angeles Basin provides the classic case history of what can happen when this possibility was overlooked.

The general population is naturally exposed to hazards which are, and can be, neither clearly defined nor precisely controlled. A study of asthma and air pollution in Yokkaichi, Japan, showed that environmental factors associated with asthmatic paroxysms included the passage of cold fronts, allergens (including mites and certain kinds of mold spores) as well as air pollution. In another study,[5] 445 children from a "highly polluted" area were compared with 415 children from a "lightly polluted" one, with the results shown in Table 2. In addition to the specifically pathological symptoms, there was some impairment of the sense of smell also.

Thus it is clear that even where the medico-hygienic hazard from known smelly emissions can be estimated with some assurance, it is not so easy to be sure of the situation as it relates to unknown or unidentified emissions, of whose existence there can be little doubt. Recently, for example, chemical analyses of the H_2S content of the flue gases from two Swedish pulp mills were compared with actual odor thresholds. At one of the mills, low hydrogen sulfide concentrations gave low odor intensity, but at the other the odor intensities were high even at low H_2S levels. This result was taken as showing that, "other unknown substances can be of importance regarding odor strength . . ."[1]

Evidently the days when the atmosphere could be used as a sewer are nearly over and the things society allows to be discharged into it will be ever more closely restricted. A normal adult not engaged in heavy physical exertion may, in the course of a day, take in about 1 kg (dry weight) of food, perhaps 3 kg of water, and, in the same interval, more than 15 kg of air will pass into and out of his lungs. At a pollutant concentration of 1 ppm, this would mean a daily dosing with 10 mg of the pollutant. Those who regard the fluoridation of water supplies and other forms of mass medication as unacceptable might do well to transfer some part of their efforts to protecting what comes in through our noses.

REFERENCES

1. **Cederlof, R., Friberg, L., Jonsson, E., Kaij, L., and Lindvall, T.**, Studies of annoyance connected with offensive smells from a sulfate cellulose factory, *Nord. Hyg. Tidskr. (Stockholm)*, 45, 39, 1964.
2. **Lindvall, T.**, On sensory evaluation of odorous air pollutant intensities, *Nord. Hyg. Tidskr. Suppl. (Stockholm)*, 2, 4, 1970.

3. **Lindvall, T.**, Monitoring odorous air pollution in the field with human observers, *Ann. N.Y. Acad. Sci.*, 237, 247, 1974.
4. **Prokop, W. H.**, Status of regulations for source emission and ambient odors, *Ann. N.Y. Acad. Sci.*, 237, 288, 1974.
5. **Shimada, T. and Iizumi, O.**, Survey of the effect of air pollution on school children in the Fuji area, Shizuoka prefecture, *Jpn. J. Otorhinolaryngol. Soc.*, 77, 72, 1974.
6. **Sandage, C.**, Tolerance Criteria for Continuous Inhalation Exposure to Toxic Material. I. Effects on Animals of 90-Day Exposure to Phenol, CCl_4, and a Mixture of Indole, Skatole, H_2S, and Methyl Mercaptan, Office Tech. Serv., AD 268, 783, U.S. Department of Commerce, Washington, D.C., 1961.
7. **Wohlers, H. C.**, Odor intensity and odor travel from industrial sources, *Int. J. Air Water Pollut.*, 7, 71, 1963.
8. **Stevens, Sir James Fitzjames**, *A General View of the Criminal Law of England*, London, 1890, 106.

Chapter 35

ODOR CONTROL

The undesirable tendency of the age towards premature specialization and undue segregation of closely interrelated branches of science calls for an insistence, particuarly in introductory courses, upon the fundamental unity of science and upon the judicious correlation of academic and technical aspects with considerations of general economic significance.

J. Read

Odor control usually means modifying or eliminating bad smells. Imparting desirable odors is called perfuming. Desirable odors need not necessarily be pleasant: it is customary to add distinctive and strong-smelling odorants to natural gas to give warning in case of leaks. There is a multitude of articles and books on the subject of odor control so that it would be invidious to recapitulate or condense them into one short chapter. What can be done is to see how far the material in Part I of this book is relevant to matters of practical application.

The first thing to dispose of is the alleged phenomenon of odor cancellation. Long ago, Zwaardemaker[1] reported that by presenting one specific odor at a precisely controlled concentration to one nostril, and a different specific odor to the other nostril, again at a precisely controlled concentration, the result was odor cancellation, by which he meant that no smell sensation was perceived by the Central Nervous System (CNS). This rather special set of conditions has been made the basis of advertised claims that industrial and other odors can be cancelled by broadcasting other smelly substances in uncontrolled amounts for simultaneous presentation to both nostrils. No experimental data has ever been published to demonstrate the reality of such an effect. It has, however, been claimed that the vibrational theory makes the phenomenon plausible by analogy with the interference effects observed with light rays. This is, of course, sheer nonsense. When light rays interfere at one place they *must* reinforce at another simply because energy must be conserved!

In some cases, the odor cancelling preparation incorporates a substance which has the property of temporarily inactivating the sensory receptors in the nose. At one time, a 40% solution of formaldehyde was sold for the purpose, dressed up with a trace of chlorophyll to add glamor. Formalin is, of course, not the only substance which can be used as an embalming fluid.

It seems evident that any preparation which reduces the sensitivity of the nose or otherwise interferes with nasal functioning must be unacceptable for a variety of reasons.

As an alternative to odor cancelling, reodorization is somewhat more acceptable for some purposes but probably not all. The term "odor masking" is sometimes used interchangeably with "reodorization" but should be avoided for the reason that the word has also been used to describe products whose real function is to impair olfactory sensitivity.

There is substantial evidence that the perceived character of a smell can be changed by adding various components to it, so that it is entirely possible for an objectionable odor to be made less offensive by covering it with suitably chosen additives. Indeed, it is known that the more successful perfume formulations for personal adornment necessarily include one or more ingredients which, by themselves, would be repulsive or even disgusting. In small amounts, they give character to the blend in which they are incorporated.

Given that reodorization is a real phenomenon, the way the effect is exploited is a matter of ethics. At the most immediate personal level, perfumes may be used either as an alternative or as a supplement to personal cleanliness. On a larger scale, cosmetic perfumes may be used in factories, offices, or eating places to an extent that is justified by the amount of prior odor control they are designed to supplement. At its best, the situation would be comparable to the use of background music as cosmetic sound to cover the noise from an air-conditioning system or the hum of fluorescent lights.

What is also legitimate is the use of a perfume to cover an odor that will be perceived only temporarily in any event. For example, following a fire in a department store or warehouse, some upholstered furniture may be quite undamaged apart from having taken up a smoky smell. This would be objectionable in a sales room stocked with such furniture, unless covered, but the covering need not last more than a few days or weeks from the time the goods are sold inasmuch as they will by then have taken up the characteristic smell pattern of their new surroundings. The same is probably true of odorous substances added to paint to make it less objectionable when freshly applied.

What is not legitimate is the broadcast release of reodorizing chemicals from the rooftop of a factory where an offensive trade is being carried on. Not only it is usually ineffective as a remedial measure, but it amounts to mass medication of everyone in the vicinity and, moreover, medication with substances of undisclosed composition whose physiological or pathogenic properties are unknown.

The use of reodorizing perfumes is fully legitimate only when every known and available means has been used to eliminate or minimize the smells they are intended to correct. Malodors are inseparable from many human activities, but they can often be kept from escaping into the environment by modifying the process, or by absorbing them, or by chemically destroying them. A rather elementary example of containment is the addition of powdered gypsum (calcium sulfate) to outdoor privies or latrines in order that its mildly acid properties will neutralize and render nonvolatile the ammonia and other basic nitrogen compounds which are responsible for most of the offensive smell.

Probably the most effective of all methods of outright chemical destruction is by burning. At the industrial level, this is often achieved simply by means of ducts to convey malodorous exhaust gases into the primary air supply of a furnace or boiler.

Chemical destruction by such substances as potassium permanganate, chlorine, or ozone can be applied in special circumstances, but is likely to be costly and to give rise to sometimes surprising side effects, but it can, on occasion, be extremely valuable. Thus, for example, when I was asked what could be done for a house where a skunk had crawled into the hot air heating system, fired both barrels and then died, the answer was: (1) remove the corpse; (2) take everyone out of the house for a full day, close all the doors and windows, and leave an ozone generator running inside; (3) mount guard continuously to make sure no one entered during the treatment because ozone is *toxic*! At the end of the day the doors and windows were opened from the outside and the house thoroughly ventilated before allowing anyone to enter it. A once-only application of ozone like this was legitimate. The use of continuous ozonizers in factories, offices, or aircraft is not legitimate because, quite apart from its inherent toxicity, its function is to act on the sensory receptors as much as on the odorous molecules themselves.

Incidentally, ozone will *not* oxidize carbon monoxide in the atmosphere of automobile repair shops.

There is, finally, the matter of controlling personal body odors. Some of these are a result of medical conditions and call for specialized treatment. Otherwise, given an adequate standard of living, this is mainly a matter of cleanliness and the avoidance of emanations from exotic foods or exceptionally vile cigars or pipes. If the advertisers are to be believed, it may be necessary to supplement these with antiperspirants (which are nothing more than tanning agents) and also with perfumes and lotions of various sorts. The degree to which they are called for or are acceptable will vary with the locality and the prevailing mores. The nose is, as always, a channel of communication so that the presence of a smell, or the absence of one, is telling us something if we are able to pay attention.

REFERENCES

1. Zwaardemaker, H., *Die Physiologic des Geruchs* Engelmann, 1895.

Chapter 36

SMELL PSYCHOLOGY

Whoso is wise will ponder these things.

Psalms: 107, 43.

Psychologists who are interested in smells and smelling usually approach the matter in one or the other of two ways. In one, they function as pure psychologists and describe (and if possible measure) the sensations but without attempting to correlate the responses with any attributes of the stimulus other than its name and, on occasion, its concentration. They record the subjects' verbal responses, and at times such objectively measurable reactions as the electrical conductivity of the skin, or the electroencephalogram, or the pulse rate, or they estimate the amount of anger or disgust or curiosity or pleasure the stimulus elicits. The results of such studies are of necessity almost wholly descriptive. For example, they tell us that people can discriminate between odors very much better than they can estimate odor intensity, and that children 3 to 5 years old find the odor of sweat or feces rather pleasant while older children find them repulsive. Other experiments show that estimates of pleasantness or unpleasantness may be conditioned by past experience, so that a relatively neutral sensation will be classed as pleasant if it comes at the end of a series of ugly ones, and as somewhat unpleasant if it comes after a set of generally agreeable ones.

Apart from these personal predispositions, there may be some degree of genetic conditioning. In one type of experiment, subjects were presented with pairs of stimuli and asked to estimate how far they were apart on some sort of mental scale of pleasantness/unpleasantness or floralness/unfloralness, or sameness/difference. The numerical data obtained in this way were then analyzed statistically and a multi-dimensional scaling of the sensations was arrived at. In at least one such study, based on estimates of pleasantness/unpleasantness, compounds whose odors were on the whole rated pleasant were likely to be in the general class of things which (1) occur in nature, and (2) are safe to eat. Things which were rated unpleasant were also in the class of things which occur in nature, but by and large they were things it is not safe to eat. Things more or less in the middle, such as alcohol, were things not found in nature and for which natural selection would not have resulted in a predisposition to like or dislike. It seems clear that individuals who eat the wrong things are less likely to become ancestors than those who choose the right ones.

In still other types of purely psychological experiment, the effects of a preexisting emotional stress or of the sociological background have been studied with interesting conclusions about such things as the public's reaction to industrial pollution of the atmosphere. In a somewhat different context, perfumes can be used to identify a product line or to act as hidden persuaders to make one product preferred over another that may be identical apart from the added scent. The classical example of this was an occasion when boxes of nylon stockings were set out for sale on an open counter. The stockings were identical except that the contents of half the boxes were perfumed and the others were not. The perfumed ones outsold the others three to one! But on another occasion, when customers complained about the bad smell of a lawn fertilizer, they refused to buy an equally good but odorless product "because it didn't smell like fertilizer".

A different sort of possibility was exploited when perfumes were sprayed into workrooms with a resulting increase in the output of finished products. In a boot-and-shoe factory watermint was used, verbena in an animal glue works, and lavender in a chemical factory.

Evidently there is an interesting field of investigation waiting to be exploited by the type of person who is more interested in getting things done than in finding things out.

For the latter sort of person the more interesting psychological approaches are those that may shed some light on the way the human brain processes information and ideas.

Language is conspicuously deficient in words relating to smell perceptions, and it would be interesting to know why. Most of the words we use are carried over from other kinds of sensation, so that one odor may be described as bright or clean and another as heavy and still another as sour. At other times an odor will be described on the basis of its association with a particular source. It may be fishy, or violet, or smoky, or like sandalwood. It would be an interesting part of social anthropology to make cross-cultural comparisons of odor preferences and odor-descriptive words.

Evidently, this aspect of the matter depends upon the mind's ability to register a present impression and to compare it with impressions or patterns stored in the memory. As is well-known, there are two apparently different but not wholly separate memory functions, known as the short- and the long-term memories. Short-term memory is called upon when two stimuli are presented and the subject is asked to smell first one and then the other and to make some sort of pronouncement about their resemblance, or intensity, or degree of pleasantness. Long-term memory is called upon when the subject is asked to find a descriptive word to associate with the sensation, and here the evocativeness of the odor is predominant. It is well-known that odors may often tend to evoke a tremendous emotional response if they are associated in the memory with some event or situation which has been important in the experience of the individual.

This, at times overriding property of the sensation can be important in more than one way. An airline pilot while landing a four-engined jet carrying some hundreds of passengers is subject to a tremendous sensory overload. His eyes are watching the ground and his instruments, his ears are listening to the flight controller, his hands are feeling the controls, and as an experienced pilot, he is drawing constantly on his memory. Yet, despite this tangle of sensory inputs, the smell of smoke in the cockpit will start alarm bells ringing in his mind because his sense of smell has priority and a direct entry into his conscious levels.

In Chapter 2 the ability of the nose to make fine discriminations between sensations provided the basis for an entire theory of the stimulatory process. Discrimination experiments necessarily call for precise specifications of the stimuli and are therefore not purely psychological. Odor recognition falls within the realm of more or less pure psychology insofar as it depends upon the ability to draw upon patterns stored in the memory. Such matters as the effect of expectation on the accuracy of an odor judgment can be studied with interesting results when, for example, identical odorants are presented on differently colored paper strips.

The ability of the subject to direct his attention to various parts of the overall sensation can be studied by presenting odors in series and then in parallel, and it may well be that some light will eventually be shed on the way the machinery of the mind works. The fact, already mentioned, that odors evoke such vivid recollections of sights, sounds, or events would seem to put the olfactory sense near the hub of the wheel.

An olfactory abnormality, known as an "olfactory paramnesia" is manifested by an inability or difficulty in correctly naming or associating a previously experienced odor sensation. Some of these are apparently innate in certain persons, but others can be induced experimentally by suggestion while the subject is under hypnosis.

The phenomenon of cross-adaptation or selective fatiguing of the sense by exposure to one stimulus before presentation of a second has not been very rewarding as long as it was taken to be a purely psychological phenomenon, but taken in conjunction with the vibrational characteristics of the respective stimuli it may well prove illuminating in ways to be considered in Chapter 46.

The foregoing matters are not intended to be an exhaustive review of the strictly psychological ramifications of osphresiology. Nor was it my object in preparing this section of the book to compile a definitive reference list. With new papers in the field of smells and smelling appearing in the technical literature at a rate of 50 to 100 per month, only a mindless computer could abstract and summarize them all. More important, it would reveal little or nothing in the way of a common thread to give them coherence like the beads on a necklace. What I have hoped to do in Part II is to suggest to those working in other disciplines ways in which their fields might be broadened and their expertise sharpened by taking olfactory phenomena into account.

REFERENCES

1. **Achelis, J. D.**, Studies of Olfaction, *Arch. Gesommte Psychol.*, 71, 273, 1929.
2. **Austrian, J. A., Van Bommel C., and Schwartz, N.**, Influence of color in the environment on perceived fragrance qualities, *J. Soc. Cosmet. Chem.*, 15(9), 465, 1964.
3. **Bannister, H. and Zangwill, O. L.**, Experimentally induced olfactory paramnesias, *Br. J. Psychol.*, 32, 155, 1941.
4. **Cederlöf, R., Friberg, L., Jonsson, E., Kaif, L., and Lindvall, T.**, Studies of annoyance connected with offensive smell from a sulfate cellulose factory, *Nord. Hyg. Tidskr. (Stockholm)*, 45, 39, 1964.
5. **Clarenburg, L. A.**, Penalization of the environment due to stench. A study of the perception of odorous air pollution by the population, *Atmos. Environ.*, 7, 333, 1973.
6. **Engen, T., Lipsett, L. P., and Kaye, H.**, Olfactory response and adaptation in the human neonate, *J. Comp. Physiol. Psychol.*, 56, 73, 1963.
7. **Engen, T.**, The effect of expectation on judgments of odor, *Acta Psychol.*, 36, 450, 1972.
8. **Engen, T., Kuisma, J. E., and Eimas, P. D.**, Short-term memory of odors, *J. Exp. Psychol.*, 99, 222, 1973.
9. **Engen, T. and Ross, B. M.**, long-term memory of odors with and without verbal descriptions, *J. Exp. Psychol.*, 100, 221, 1973.
10. **Foster, D.**, Odors in series and parallel, *Proc. Sci. Sect. Toilet Goods Assoc.*, 39, 1, 1963.
11. **Jonsson, E.**, Annoyance reactions to external environmental factors in different sociological groups, *Acta Sociol. (Copenhagen)*, 7, 229, 1964.
12. **Mitchell, M. A., Konigsbacher, K. S., and Edman, W. W.**, The importance of odor as a nonfunctional component, or odor — a tool of marketing, *Ann. N.Y. Acad. Sci.*, 116, Art. 2, 685, 1964.
13. **Moncrieff, R. W.**, Changes with age in olfactory preferences, *France Parfums*, 8(46), 397, 1965.
14. **Rovesti, P.**, Some psycho-physiological and social aspects of the use of perfumes, *Soap Perfum. Cosmet.*, 38, 143, 1965.
15. **Spealman, C. R.**, Odors, odorants and deodorants in aviation, *Ann. N.Y. Acad. Sci.*, 58, Art. 2, 40, 1954.
16. **Woskow, M. H.**, Multi-dimensional scaling of odors, in *Proc. NATO Summer School, Istanbul, 1966*, Robert College Research Center, Bebek, Istanbul, 1968.

Chapter 37

SMELL PSYCHOLOGY — II

Kepler's marvellous achievement is a particularly fine example of the truth that knowledge cannot spring from experience alone but only from the comparisons of the intellect with observed fact.

A. Einstein

The essence of scientific research is the quest for correlations. Merely to state the length of a certain piece of string is totally uninteresting, but if it was used to wrap a book or catch a fish it can be correlated with the lengths of other bits of string used for similar purposes and some usable (if not especially useful) conclusions drawn. In the last chapter, correlations were sought between events or processes all of which were essentially within the minds of essentially normal persons. At least two other kinds of correlation can be looked for between psychological and overtly physiological responses to odorous stimuli. Some, relating to what have been called External Chemical Messengers (ECMs) have already been touched on in Chapter 28, and it is sufficient here to remember that abnormalities or disorders of perception may at times serve as tools in the diagnosis of anxiety states or as correlates with such things as neurodermatitis or asthma. In one reported study, patients were tested with three kinds of sensory stimulation: cutaneous, muscular, and olfactory. Following each type, the subject was asked to make a drawing of anything that came to mind and then to describe and associate the drawing. Cutaneous stimulation produced the strongest emotional responses in dermatitis patients and olfactory stimulation in asthmatics. The subjects were usually quite unconscious of a connection between the stimulus and the drawings, so that the linkage seems to have been at a preconscious level.

Passing from such effects which are perceived essentially subjectively *by the observing psychologist*, one meets intermediate cases where the attempt is made to correlate photographically recorded facial expressions of the subject with the type of odor presented. The results were interpreted as stimulus-dependent communication signals and, moreover, signals which are largely innate inasmuch as newborn and not-yet-fed infants tend to conform to the same patterns.

Experiments such as these in which information received by the subject from an olfactory input is conveyed to the observer by visual means are not essentially different from those in which the subject-to-observer communication is verbal and of the type considered in the previous chapter. It is worth remembering, however, that the olfactory sense is part of a communication channel between persons who need not be in the position of investigator and subject. Thus, it has been reported that if the mother or nurse regularly wears the same perfume, babies showed a relaxed behavior, an increase in cooing, etc. when the odor was present regardless of who was wearing it. This is entirely analogous to the role of maternal odors in small animals or their ability to recognize alien members of their own species or to perceive a difference between stressed and unstressed individuals. Closely related to these are such phenomena as the fright reaction which causes a school of fish to scatter, or the home-stream odor which enables mature salmon to return to the same ground where they were spawned. These effects may be genetically fixed or the result of imprinting — or to a combination of both if imprintability is inherited.

The role of olfaction in interpersonal relations — which may be called social olfaction — is important whether the signal is perceived consciously or whether it operates below the level of conscious perception. The role of pheromones in the animal and insect worlds has been studied intensively ever since the word pheromone was invented, and so it has been natural to look for possible pheromone effects in humans also. A pheromone is defined as a chemical released by one individual which produces a response in another individual of the same species. Based on the type of response, they can be subdivided into two classes. Primer pheromones affect the physiology of the receiver as manifested by such things as the Bruce Effect (an olfactory pregnancy block in newly mated female mice who fail to conceive if exposed to the odor of alien males), or the remarkable physical and behavioral changes in desert locusts when they pass from the solitary to the migratory phase as a result of becoming crowded together. At the human level, perhaps the most striking example of a primer pheromone being at work is McClintock's[13] report of menstrual synchronization in pairs of women living together in a college dormitory.

The possible role of releaser pheromones (pheromones which release a behavioral response from a physical or physiological situation) is more accessible to study, and possible traces of their operation have been reported in such things as the fact that women are more able than men to detect certain male chemical emanations such as boar taint in pork or the differential effects on men and women of exposure to sex-related odors prior to undertaking some kind of psychological test.

It has to be remembered that where drugs operate at a level of milligrams, pheromones operate at the level of molecules, thanks to their role as conveyers of information rather than as building blocks. In the lower animals their effects are usually quite explicit, whereas in man they are more often diluted because of the larger variety of signal systems and the greater override from learned or conditioned behavior.

Turning now from the sort of olfactory response that can be registered without instrumental devices, we come to the more precise and quantifiable things that can be recorded by measuring such things as the rate of heartbeats, the electrical conductivity of the skin, or the electroencephalogram. These experiments have repeatedly confirmed the extreme sensitivity of the olfactory sense when, for example, it turns out that there are unmistakable changes in various bodily responses before the subject is consciously aware of the presence of a smell. The experiments appear to have left open such questions as, for example, whether pleasant odors have more effect on the EEG and repulsive ones on the electrical resistance of the skin (or vice versa), or whether sex-related compounds operate primarily on the heart rate. It is not unreasonable to look for specific effects of this sort inasmuch as the smell of a well-cooked meal at the right time and place will produce a mouth-watering response which the smell of pine needles or cigar smoke would never provoke.

Probably the most unequivocal result of these experiments is an objective demonstration of the essential difference between habituation and sensory fatigue. Sensory fatigue is manifested by a reduced ability to perceive a certain sensation after a more or less prolonged exposure to a higher level of the stimulus in question. It seems evident that it results from a partial exhaustion of the receptor mechanisms. Habituation is different. It is simply the mind's ability to disregard a background noise, or visible object, or a smell that is irrelevant to the mind's preoccupation of the moment. The background stimulus can be perceived instantly if attention is drawn to it. Habituation, then, is closely related to the same mental mechanisms that allow one to direct one's attention to the various notes (camphoraceous, spicy, etc.) present in a complex odor sensation.

It is possible that at least some of the reportedly subliminal odor responses take place not when the odor is actually imperceptible but rather when the subject fails to perceive it because the attention is directed elsewhere. While the smell of smoke may have an overriding significance to an airline pilot, the smell of food might not. The perfume worn by a stewardess would probably fall somewhere between these extremes. A person's mental or emotional state is known to affect the olfactory threshold, as measured by the stimulus concentration needed to produce a response at the conscious level.

REFERENCES

1. **Asaka, J.**, The studies on the objective olfactory test by galvanic skin response reflex, *J. Otorhinolaryngol. Soc. Jpn.*, 68, 100, 1965.
2. **Bushteuva, K. A., Polezhaev, E. F., and Semenenko, A. D.**, The influence of subliminal olfactory stimulation on reflex activity (transl.), *Fiziol. Zhur. SSSR*, 46, 533, 1960.
3. **Carr, W. J., Roth, P., and Amore, M.**, Responses of male mice to odors from stressed vs. nonstressed males and females, *Psychon. Sci. Sect. Anim. Physiol. Psychol.*, 25, 275, 1971.
4. **Comfort, A.**, Likelihood of human pheromones, *Nature*, 230, 432, 1971.
5. **Cowley, J. J.**, The Effect of Exposure to Odorous Compounds (Pheromones) on an Assessment of People Test, *1st ECRO Congr.*, Paris, July 3 to 5, 1974.
6. **Frosecchi, M. and Digiesi, V.**, Influence of odorous stimuli on peripheral circulation in man, *Rass. Neurol. Veg.*, 18, 100, 1964.
7. **Engen, T., Cain, W. S., and Rovee, C. K.**, Comparison of olfaction in the newborn infant and the adult human observer, in *Proc. NATO Summer School, Istanbul, 1966,* Robert College, Bebek, Istanbul, 1968.
8. **Hasler, A. D.**, Guideposts of migrating fishes, *Science*, 132, 785, 1960.
9. **Hemming, C. F.**, The Locust Menace, *Anti Locust Research Center*, London, 1968, 31.
10. **Kepecs, J. G., Robin, M., and Munro, C.**, Responses to sensory stimulation in certain psychosomatic disorders, *Psychosomat. Med.*, 20, 351, 1958.
11. **Kloek, J.**, Psychological and neurophysiological aspects of flavor compounds, *Bibl. Nutr. Dieta*, 9, 105, 1967.
12. **Kottmeyer, G.**, Changes in blood supply produced by olfactory stimuli, *Zentralbl. Hals. Nas. Ohrenheilk*, 58, 122, 1957.
13. **McClintock, M. K.**, Menstrual synchrony and suppression, *Nature (London)*, 229, 244, 1971.
14. **Neuhaus, W. and Goldenberg, D. M.**, Some vegetative responses to olfactory stimuli in man, *Experentia*, 22, 174, 1966.
15. **Parkes, A. S. and Bruce, H. M.**, Olfactory stimulation in mammalian reproduction, *Science*, 134, 1049, 1961.
16. **Pfeiffer, W. and Lemke, L.**, On the isolation and identification of the alarm substance from the skin of the European minnow, *Phoxinus phoxinus* L., *J. Comp. Physiol.*, 82, 407, 1973.
17. **Rovee, C. K., Harris, S. L., and Yopp, R.**, Olfactory thresholds and level of anxiety, *Bull. Psychonom. Soc.*, 2, 76, 1973.
18. **Steiner, J. E.**, Discussion paper: innate, discriminative human facial expressions to taste and smell stimulation, *Ann. N.Y. Acad. Sci.*, 237, 229, 1974.
19. **Veylon, R.**, Biochemical regulation of social behavior. Pheromones, *Nouv. Presse Med.*, 1, 2993, 1972.

Chapter 38

SMELL PSYCHOLOGY — III

Truth is rarely pure, and never simple.

Oscar Wilde

Inasmuch as most smell stimuli are chemical mixtures, the psychological matters discussed in the preceding chapters were considered without much reference to the general theory of olfaction developed in Part I. It remains to be seen how some of the avenues it opens may lead to a possibly better understanding of various phenomena.

Moncrieff,[5] for example, reported that there are noticeable differences between the psychogalvanic reflexes of different people, and went on to suggest that the method is not likely to be useful in classifying odors but may be useful in classifying people. As regards his first point, it is true that psychogalvanic reponses are unlikely to make fine discriminations between such things as spearmint and peppermint, and can probably tell us very little about the floralness of a particular chemical. On the other hand, these responses may tell us something — and perhaps a great deal — about the messages being received in the Central Nervous System (CNS). If, for example, it can be shown that unpleasant or repulsive odors produce a pronounced alteration in the electrical resistance of the skin, but leave the heart or respiratory rates largely unaffected, then we would have an objective method of determining the kind of information the signal carries, independently of the words the subject uses to describe it.

Focusing our attention on the message rather than on the messenger has interesting effects on the emphasis in our thinking. Thus, for example, J. Kloek[4] studied the odors of some steroid sex-hormones and their metabolites and remarked that, "Steroids are biologically important substances, closely connected with sexuality; in animals as well as in plants they are much more generally distributed than macrocyclic compounds like musk. Moreover, the excretion of musk must be considered as a secondary sex characteristic . . . Therefore we better can say: *musk and other sex-attractants do have the smell of steroids; the steroids have to be considered as the substances which take the central position as sex-attraction agencies.*"[4]* This implies that the ability to secrete the chemical, muscone, (Chapter 4) has been evolved by natural selection from a metabolic accident which produced a chemical with no inherent physiological function in itself but which can masquerade as one with such a function.

In Chapter 21 it was pointed out that the chemicals used as sex attractants by insects have no across-the-board similarity because they are not involved *as chemicals* in the reproductive process. Steroids are so involved in the chemistry of reproduction and the fact that they are also odorous is essentially coincidental to the biological but not to the behavioral aspect of the matter. To the extent that musk has a similar but stronger odor of the same sort, it helps to ensure the survival of the species through its function as a pheromone.

A further interesting aspect of Kloek's paper[4] is in the words, other than "musk", used by both students and professional perfumers to describe various steroids. Naturally, many were described as having a "general animal" smell, or as being "like perspiration", but in addition such words as "wood", "sandalwood", "cedar", "dead leaves", "amber", and, most interestingly, "like incense", turned up fairly often. The importance attached to the delicate scents of some, but not all, kinds of trees in many religious ceremonies takes on an added interest in relation to the role of olfaction in social anthropology.

* Italics in the original.

Information relating to sex is, of course, not the only kind of message that may be carried by a smell. Emotions of alarm, fury, joy, sorrow, anxiety, and perhaps even hunger may be involved in various aspects of social life and may play a part in riotous outbreaks, contagious panic, and other aspects of crowd behavior. Some possible ways to study these phenomena are discussed in Chapter 50.

In Part I, molecular vibration-patterns were correlated with words used to describe human sensations and also with the behavioral responses of insects. If human heart rates, or encephalograms, or salivation can be correlated with the inputs derived from certain chemicals, our comprehension of both sides of the stimulus-response equation may well be broadened.

There is, however, a complication in the exploration of this pathway. What we perceive as a smell sensation very often represents a fusion of information carried by the olfactory nerves *plus* information brought by a quite distinct nerve network, known as the trigeminal system. This latter apparatus picks up its messages from terminals in the eyes as well as in the nose. Its most familiar manifestation is the stinging sensation in the eyes as a result of exposure to ammonia and other irritating vapors. Its principal function is to transmit a rather generalized pain sensation, and it is somewhat analogous to the common chemical sense mentioned in Chapter I which gives an undifferentiated pain sensation when salt enters a wound. In relation to olfactory inputs its role is to impart a pungent note to such stimuli as vinegar or oil of cloves. At the level of conscious awareness, the fusion of the two kinds of sensory input is so intimate that it is usually difficult to separate their separate contributions to the perceived smell.

One method of doing so has been reported by D. Smith et al.[6] They used measurements in the scalp to register cerebral responses to odorous stimuli, so as to detect what they called the odorant evoked response. By selectively blocking either the olfactory or the trigeminal pathways with cocaine, they found that the evoked response was abolished if the trigeminal inputs were abolished, but not when the olfactory inputs were cut off. And so they concluded that under the experimental conditions they used, the response as registered in the scalp appeared to result entirely from stimulation of the trigeminal nerves.

Some years earlier, W. F. Allen[1] approached the same problem from a different direction, using conditioned reflexes. Using dogs, he offered olfactory stimuli simultaneously with electrical stimulation of the paw so that the animals learned to lift the paw when the olfactory stimulus was presented without the electrical one. If, subsequently, the olfactory or the trigeminal nerves were cut, the continuation of the conditioned response would depend upon whether the conditioning stimulus was purely olfactory or in some degree trigeminal also. He concluded that the odors of cloves, lavender, anise, asafetida, and benzene are truly olfactory *in the dog,* while those of camphor, eucalyptus, pyridine, butyric acid, and phenol are in part trigeminal. The same distinction probably does not hold with humans inasmuch as a trained police dog can function apparently normally while exposed to a concentration of tear gas that renders his master practically helpless.

Recently R. L. Doty[2] studied 14 human subjects who, for some reason, lacked a sense of smell. He found that more than a quarter of the 31 chemicals he used as stimuli could be detected by the trigeminal receptors in the nose. In those cases where he estimated the detection thresholds, they were higher than for normal persons by a factor of about 100. The Stevens exponents (Chapter 17) were not very different from those of normal subjects for the same stimuli. It is clear, therefore, that odor sensations as perceived at the level of conscious awareness are in many cases the integrated end-product of two distinct kinds of input. Of the two, the olfactory inputs are much more highly discriminating and the message specificity described in Chapter 2, derives almost wholly from the olfactory component.

Earlier, and again with human subjects, C. A. Elsberg[3] studied the relative roles of olfactory and trigeminal stimulation by measuring their respective fatigue effects. His experimental techniques have been largely if not entirely superceded for a variety of reasons, but his general conclusion that many odors are really combined sensations which sniffing does not separate remains unchallenged. As mentioned in Chapter 28, Elsberg's main interest was in seeing how far lesions in the relevant parts of the brain can be diagnosed by peculiarities in the patient's sense of smell. It would probably be worthwhile to retrace his steps using electrical or other measurements to determine the response to chemicals that are simply sniffed rather than by his highly artificial blast- and stream-injection procedures.

REFERENCES

1. **Allen, W. F.,** Olfactory and trigeminal conditioned reflexes in dogs, *Am. J. Physiol.,* 118, 532, 1937.
2. **Doty, R. L.,** Intranasal trigeminal detection of chemical vapors by humans, *Physiol. Behav.,* 14(6), 855, 1975.
3. **Elsberg, C. A., Levy, I., and Brewer, E. D.,** The sense of smell. VI. The trigeminal effects of odorous substances, *Bull. Neurol. Inst. N.Y.,* 4, 270, 1935.
4. **Kloek, J.,** The smell of some steroid sex-hormones and their metabolites. Reflections and experiments concerning the significance of smell for the mutual relation of the sexes, *Psychiatr. Neurol. Neurochir.,* 64, 309, 1961.
5. **Moncrieff, R. W.,** Psychogalvanic reflexes to odors, *Perf. Essent. Oil Rec.,* 54, 313, 1963.
6. **Smith, D., Allison, T., Goff, W., and Principato, J.,** Human odorant evoked responses: effects of trigeminal or olfactory deficit, *EEG Clin. Neurophysiol.,* 30(4), 313, 1971.

Chapter 39

THE RETURN OF THE SALMON

I love any discourse of rivers, and fish, and fishing.

Isaak Walton

In April, 1958, some thousands of steelhead fingerlings (*Salmo gairdnerii*) 6 to 8 in. long were released from the Alsea River fish hatchery on the coast of Oregon. Before they were let go they were marked by clipping certain nonessential fins in a special way. Five months later, on September 5, 1958, one of these fish was caught in a net off the coast of Alaska, 2000 mi away. It was then about 14 in. long. It was marked with a numbered tag and put back in the sea. Seventeen months later, on February 5, 1960, this same fish, now more than 2 ft long, turned up back at the Alsea hatchery where it had been born 2 years earlier.

A similar experiment was carried out on the other side of North America in 1938 when A. G. Huntsman[4] marked 31,359 young Atlantic Salmon (*Salmo salar*) before releasing them in the Northeast Margaree River in Nova Scotia. On June 17, 1940, one of these fish was caught, tagged and released off Bonavista, Newfoundland, and 3 months later, on September 21, 1940, it was caught by an angler on the Margaree River within 2 mi of the place it was released, and 570 mi by the shortest sea route from Bonavista.

It has long been known that various species of salmon spawn in fresh water and later go to sea for several years before returning to the fresh water to lay their eggs. What made these observations so significant was the fact that these particular fish were registered and marked before they left the home stream, reregistered and remarked far out in the ocean, and registered a third time when they were back at their starting point. Before that, it could not have been known for sure that the returning fish had ever gone farther than the estuary of their home stream during their excursion into the salt water.

That the salmon did venture out into the high sea was rendered highly probable (but not a certainty) by such experiments as the one in which 469,326 Pacific salmon fingerlings (*Oncorrhynchus nerka*) were marked before leaving a spawning ground at Cultus Lake, British Columbia, on a tributary of the Fraser River. In due course, 16,553 of them were recovered as adults, 4,995 actually in Cultus Lake and 11,558 in the approaches to it, and although traps were placed in nearby tributaries of the Fraser, practically none of the marked fish strayed into them.

Not all the experiments give results as precise as this. There is usually some straying and, indeed, both the homing of the majority and the straying of a minority are necessary to ensure the biological survival of the species.

If a fish has grown to maturity and is ready to spawn, it must have survived all the hazards of a dangerous world, and the first step to success took place when it was born in a probably favorable environment. If it can lay its eggs in identically the same environment, it may not be the best possible, but it is one that is demonstrably favorable. At the same time, some straying is desirable if new "fish runs" are to be established in new streams. This may be desirable if the way to the original home is blocked by a landslide or a power dam. A new run can be established only if a few fish are willing to try something new.

On the west coast of North America, each stream, and often each tributary, may have its own run of salmon, the timing of which can be predicted with some precision. On occasions when a run has been lost as a result of some natural or manmade calamity, a new and previously unknown run has been found somewhere else and presumed on the basis of its timing to be a transfer from the lost run.

The fact that fish spawned at a particular site time their return to it so as to arrive more or less together and thereby constitute a run also helps to ensure the survival of the species. If the fish trickled back over a whole year, or a whole summer, the population of bears and other predators would expand to take advantage of a continuous and plentiful food supply. By restricting the return to a rather short period, the phenomenon of predator satiation operates: for the time being there are too many fish for the bears to eat, and for the rest of the year there are too few to support a large population of bears. This phenomenon operates in many quite different connections. Many species of tree, for example, produce an abundance of seeds in some years and hardly any in others.

Thus the biological survival of the salmon depends upon two quite different things: the ability to make their way back to the home stream and the ability to time the return properly. While the timing of the run is probably inherited (that is, genetically controlled) its destination is probably not. If it were, the offspring of fish that stray into the wrong stream would still go back to the ancestral one and no new runs could be established. Both aspects of the matter were demonstrated in the following experiment.

In 1949, a large number of eggs of the sockeye salmon (*O. nerka*) were collected on the spawning grounds of the Horsefly River in British Columbia and transported in tanks to a fish hatchery on Horsefly Lake where they were allowed to hatch and where they spent their first year. There is a range of mountains between the Horsefly River and Horsefly Lake so the eggs were spawned in one watershed area and hatched in an entirely different one. In 1950, 94,000 fingerlings from these eggs were taken by air to the mouth of the Horsefly River where it empties into Quesnel Lake and released there after 64,500 of them had been marked. From there they made their way to the sea in the normal way.

In 1952, when some of them were due to return as 3 year olds, a close watch was kept for them at the spawning grounds on the Horsefly River and also at the hatchery on Horsefly Lake. At the spawning ground 6829 3-year-old sockeye were counted and 2228 were examined for the hatchery mark. Not one was marked. At the hatchery, 13 3-year-old sockeye turned up, 9 of them marked, and this at a place where no sockeye had previously been taken.

Since not all sockeye return after 3 years, the watch was continued into 1953. On the Horsefly River, 105,000 4-year-old fish were counted and 46,917 were checked for the hatchery mark. Only one had it. Meanwhile at the hatchery 203 marked and 66 unmarked sockeye were taken actually at the end of the outlet pipe, and 15 more marked fish were found which had spawned and died downstream from the hatchery.

The time of the new run to the hatchery coincided with the regular Horsefly River run, showing that the time of the run is probably inherited but its destination can be changed.

This experiment, and similar experiments elsewhere including some in Norway, showed that the fish did not return to the stream their parents preferred, but to the one in which they themselves were hatched and spent their impressionable years. The name "imprinting" has been given to a well-known phenomenon in which animals, birds, and fish become endowed with a lifelong trait or behavioral pattern as a result of being exposed to (or "imprinted" with) a certain environmental factor at a certain critical period which is usually rather early in their development.

This experiment is only one of several which show that the home stream migration of salmon is a result of imprinting rather than a matter of some mysterious homing instinct, but it has an added feature of profound significance. The eggs that were laid in the Horsefly River were transported by air over the mountains to the hatchery, and

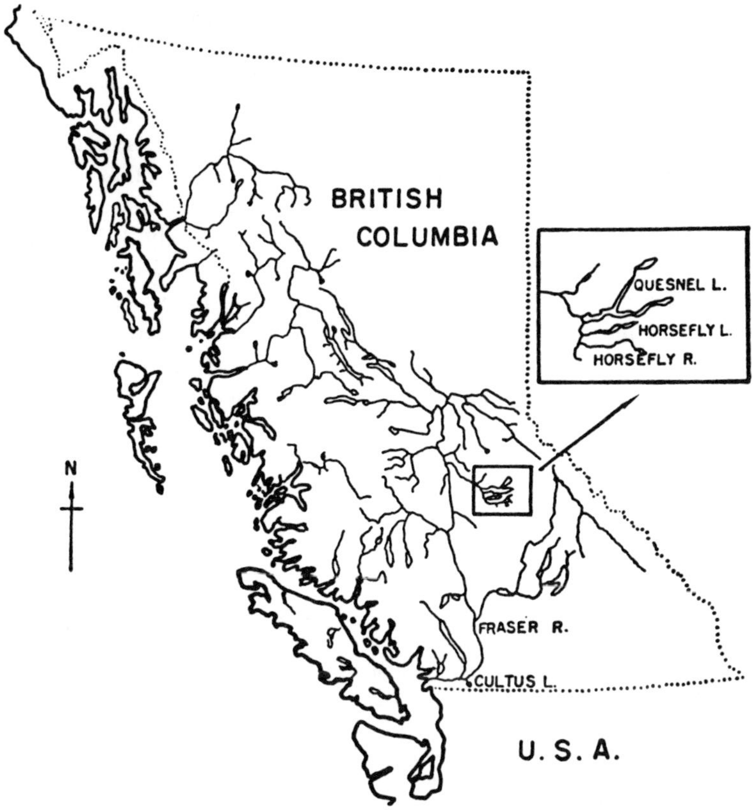

FIGURE 1. The rivers and streams shown on this map of British Columbia are the ones that support salmon runs. Streams in the northeast corner of the Province drain into the Arctic Ocean; and streams in the southeast corner that used to provide spawning grounds have been cut off from the sea by hydroelectric dams in the U.S. The northwest corner is mostly unexplored. (From Wright, R. H., *The Science of Smell,* George Allen & Unwin (Publishers) LTD., London, 1964, 59. With permission.)

the fingerlings were later flown from the hatchery to the release point on Quesnel Lake. Therefore, when these fish came back as part of the regular Horsefly River run, their destination had to be reached by a route they had never travelled before. What they remembered was not the topography of the land but the quality of the water, and this memory took them right to the effluent pipe at the hatchery.

That the specific quality of the water was its smell was shown by the following experiment by A. D. Hasler and W. J. Wisby.[6] In the state of Washington, there is a run of coho salmon (*O. kisutch*) to the Issaquah River. The fish regularly distribute themselves between spawning grounds in the main stream and in a tributary, the East Fork. In 1952, the experimenters installed a fish trap on the East Fork about a mile above the junction. There was already a trap on the main stream upriver from the fork. The procedure was to capture fish which had followed either the main stream or the East Fork, mark them with tags and put them back in the main river, downstream from the fork. Before releasing them, some had their olfactory organs plugged with vaseline, or benzocaine ointment, or cotton, or a combination of these, while others were returned to the water with their smelling apparatus intact.

Altogether, 302 fish were moved back downstream, 226 from the main Issaquah and 76 from the East Fork. Of the 302, 153 had their olfactory end-organs blocked and 149 served as controls. Not all the fish were recaptured, but those that were told a very consistent story. Of the controls from the main stream, 46 were retaken and all had repeated their original choice of streams. Of the controls from the East Fork, 27 were retaken, 19 back in the East Fork and 8 in the main Issaquah. The ability of the fish to repeat the first choice was good but not perfect.

The fish whose noses were stopped up distributed themselves in an essentially random fashion between the two streams if we allow for the fact that the main Issaquah is much larger than the East Fork. The original catch from the Issaquah was 226 compared with 76 from the East Fork: a ratio of 3:1 approximately. Fifty-one plugged fish originally from the Issaquah were recaptured, 39 there and 12 in the East Fork, again a ratio of 3:1 very nearly. Nineteen East Fork fish with plugs were retaken, 16 in the main stream and 3 in the fork, for a ratio of 5:1 which, considering the small numbers, is not significantly different from the ratio of 3:1. Denied the ability to follow their noses the plugged fish could do no better than follow the crowd.

REFERENCES

1. **Allen, G. H.**, Ocean migration and distribution of fin-marked coho salmon, *J. Fish. Res. Bd. Can.*, 23(7), 1043, 1966.
2. **Cragie, E. H.**, A preliminary experiment on the relation of the olfactory sense to the migration of sockeye salmon, *Trans. Roy. Soc. Can.*, (V), Ser., 3, 20, 215, 1926.
3. **Forester, R. E.**, The return from the sea of sockeye salmon, with special reference to percentage survival, sex proportion and progress of migration, *J. Biol. Bd. Can.*, 3(1), 26, 1936.
4. **Huntsman, A. G.**, Return of marked salmon from a distant place, *Science*, 95, 381, 1942.
5. Annual Report, International Pacific Salmon Fisheries Commission, New Westminster, B. C., Canada, 1953, 22.
6. **Wisby, W. J. and Hasler, A. D.**, Effect of olfactory occlusion on migrating silver salmon, *J. Fish. Res. Bd. Can.*, 11, 472, 1954.

The following references are included for the benefit of those who may wish to learn something of the nonolfactory processes by which fish are able to navigate in the open ocean.

7. **Bennett, M. V. L.**, Electrolocation in fish, in Orientation: sensory basis, *Ann. N.Y. Acad. Sci.*, 188, 242, 1971.
8. **Kalmijn, Ad. J.**, Electro-Orientation in Sharks and Rays: Theory and Experimental Evidence, NTIS AD-776 253, U.S. Department of Commerce, Washington, D.C., 1973.
9. **Leggett, W. C.**, The ecology of fish migrations, *Ann. Rev. Ecol. Syst.*, 8, 285, 1977.
10. **Patten, B. C.**, The rational decision process in salmon migration, *J. Cons. Int. Explor. Mer.*, 28(3), 410, 1963.
11. **Rommel, S. A. and McCleave, J. D.**, Prediction of oceanic electric fields in relation to fish migration, *J. Cons. Int. Explor. Mer.*, 35(1), 27, 1973.
12. **Royce, W. F., Smith, L. S., and Hartt, A. C.**, Models of oceanic migrations of pacific salmon and comments on guidance mechanisms, *U.S. Fish. Bull.*, 66(3), 441, 1968.
13. **Saila, S. B. and Shappy, R. A.**, Random movement and orientation in salmon migration, *J. Cons. Int. Explor. Mer.*, 28(1), 153, 1963.
14. **Walcott, C., Gould, J. L., and Kirschvink, J. L.**, Pigeons have magnets, *Science*, 205, 1027, 1979.

Chapter 40

SOCIAL INTERACTIONS OF FISH

Another law there is, which toucheth them as they are sociable parts united into one body; a law which bindeth them each to serve unto other's good.

Hooker (Eccles. Polity)

The ability of salmon to synchronize their return and smell their way upstream to the ancestral sprawning ground is a spectacular and economically important manifestation of social interaction in fish, but it is not by any means the only one. Homing responses of newts also depend upon olfactory cues, as shown by the fact that they no longer respond if the olfactory epithelium is treated with formaldehyde, and moss-head sculpin (*Clinocottus globiceps*) return to specific tide pools if they are displaced.

The information carried by an odor can influence many kinds of fish behavior, just as it can affect the behavior of an insect or a man. In 1938, K. von Frisch[4] discovered that if a minnow which had been accidentally injured or surgically operated on was introduced into an aquarium where a school of minnows was peacefully feeding, there was an almost instantaneous fright reaction. The minnows flee from their food, crowd together, and retreat again at the smallest movement or without any obvious reason. They are evidently in a highly nervous state and may not regain their confidence for many minutes or hours or even days. In the natural state this sort of communication within the group makes all the members aware of a predator almost at the instant it takes its first bite at its first victim.

W. Pfeiffer[15] has reported that the alarm signal is carried by a chemical that is produced in specialized cells in the skin. The cells do not open to the outside and so they can release their contents only when the skin is injured. As with more familiar (to us) olfactory signals, the alarm substance is effective at almost unimaginably small concentrations. The amount equivalent to 0.01 mm^2 of skin (about 0.002 mg) is sufficient to drive a school of minnows from their feeding place when added to a 14 ℓ aquarium. The fright reaction is innate and appears at a certain age regardless of prior experience. From the standpoint of evolution, the phenomenon is clearly analogous to the alarm pheromones of ants and other social insects.

Turning to a different sort of marine creature and a different sort of response, the role of sex hormones in determining the behavior of lobsters has been studied. The animals were tested in different combinations of sexes and moult stages, and the behavior was analyzed in a way to be described later, in Chapter 42. What matters here is the conclusion reached by the observers, J. Atema and D. G. Engstrom:[2]

> It seems that the lobster, as might be predicted from its anatomy, lives in a world where chemical communication is important. Our studies indicate that the reproductive behavior is largely under pheromonal control . . . Although the division between taste and smell in crustacea is poorly understood, behavioral analogies with insects and even catfishes suggest strongly that social communication is linked with the sense of smell, and feeding behavior with the sense of taste.*

At the same time these authors were careful to point out that "The sex pheromone is only one of the many social communicants that we expect in the lobster."[2]

The role of sex pheromones in other crustacea has also been established, for example by E. P. Ryan,[16] and a different but obviously related manifestation of the role of

* From Atema, J. and Engstrom, D. G., *Nature (London)*, 232, 261, 1971. With permission.

chemicals as messengers is contained in a recent report that they help to regulate the maternal care of young crayfish.

It is sometimes supposed that the olfactory sense of fishes and other nonair-breathing creatures cannot operate on the same principles as that of land animals. It is difficult to understand this objection. For reasons discussed in Chapter 19, the exterior surfaces of the receptor sites of both animals and insects are normally bathed in an aqueous film, just as are those of fishes. The two things about olfaction that distinguish it from taste and the common chemical sense are its high sensitivity and its ability to discriminate between closely similar signals, or to repond selectively to some and ignore others completely. Both are developed equally in land- and water-dwelling creatures.

To investigate sensitivity, J. R. McBride et al.[13] introduced aqueous extracts of natural foods into an aquarium holding juvenile sockeye salmon and found they could detect concentrations of food odor (whatever it was) of the order of one part in 80 million, based on the wet weight of the whole food. Again, A. D. Hasler[9] showed that a water rinse of a sprig of an aquatic plant when diluted 1 to 10,000 was detected by bluntnose minnows when further diluted in a 7 gal aquarium. The marine snail, (*Hyborhynchus notatus*) manifests a feeding response to a glycoprotein isolated from oysters at concentrations as low as 1 or 2×10^{-10} *M.* Even more extreme sensitivity was found in the European eel, *Anguilla anguilla,* L. which could be trained to differentiate between various synthetic odorous compounds such as phenylethyl alcohol or menthol. Dilutions of the pure materials ranged from 10^{-15} to 10^{-18} and it was possible to calculate that at times there was only one molecule of the odorous material present in the nostrils. At threshold levels it appeared that central olfactory stimulation would occur if two or more stimuli were received by the nose within about 3 sec.

This degree of sensitivity is inconceivable in terms of an interaction that involves chemical bonding of the stimulus to the receptor, but it is quite understandable given the kind of interaction described in Chapter 12.

On the score of discrimination, the ability of migrating salmon to perceive differences in the odors of various tributaries to a river is an obvious example. In the laboratory, yellow bullheads (*Ictalurus natalis*) which have been blinded and then trained by reward and punishment methods, could discriminate *between individual bullheads of their own species* as long as their sense of smell was intact. Thus their powers of discrimination are comparable to those of the dog which can perceive a difference in the body odors even of identical twins. This is, however, not altogether surprising given that the sense of smell appeared very early in the evolutionary process and has served a generally similar function ever since.

In passing, it would be interesting to know whether a sense of smell plays a part in the lives of plants and trees. A sex attractant chemical has been reported in females of the marine plant *Fucus serrattus* L.

The foregoing is at best a too-superficial look at the role of chemical messengers in the lives of water-dwelling creatures. It remains to mention some of the possible ways they may be turned to our account.

It is known that salmon en route to their spawning grounds exhibit an alarm response and interrupt their upstream movement if a human hand, or a bear's paw, or a piece of skin from a seal or sea lion is dipped in the water ahead of them. The mammalian skin factor as it was called prior to its identification as the amino acid, l-serine, is effective at concentrations as low as 1 part in 80,000 million (1 in 8×10^{10}) of water. It has been suggested that this substance could be used to deflect migrating salmon from unsuitable river locations or away from the unnavigable tail races of hydroelectric generators.

More generally, Hasler and Wisby[10] showed that fish could be cleared from an area without polluting it to a toxic level by the addition of less than 10^{-7} mg/ℓ of potassium phenylacetate to the water.

The same authors suggested that minnows can be trained to respond visibly to some common industrial wastes at concentrations near or below the threshold of chemical detectability. At the other extreme, Bardach et al.[3] showed that several commonly used household detergents can reduce the olfactory sensitivity of yellow bullheads and are therefore an environmentally dangerous hazard to aquatic life at levels far below those at which they become noticeably toxic.

With the growing importance of fish farming as a source of food, it is evident that chemical messengers are likely to play an ever-increasing part in the rational management of the fish.

REFERENCES

1. **Alderdice, D. F., Brett, J. R., Idler, D. R., and Fagerlund, U.,** Further Observations on Olfactory Perception in Migrating Adult Coho and Spring Salmon — Properties of the Repellent Mammalian Skin, Prog. Rep. Pac. Coast Sta. Fish. Res. Bd. Can., No. 98, 1954.
2. **Atema, J. and Engstrom, D. G.,** Sex pheromones in the lobster, *Homarus americanus*, *Nature (London)*, 232, 261, 1971.
3. **Bardach, J. E., Fujiya, M., and Holl, A.,** Detergents: effects on the chemical senses of the fish, *Ictalurus natalis*, *Science*, 148, 1605, 1965.
4. **von Frisch, K.,** An alarm substance in fish skin and its biological significance, *Z. Vgl. Physiol.*, 29, 46, 1941.
5. **Grant, D., Anderson, O., and Twitty, V.,** Homing orientation by olfaction in newts (*Taricha rivularis*), *Science*, 160, 1354, 1968.
6. **Green, J. M.,** Evidence for homing in the mosshead sculpin, (*Clinocottus globiceps*), *J. Fish. Res. Bd. Can.*, 30, 29, 1973.
7. **Gurin, S. and Carr, E. W.,** Chemoreception in *Nassarius obsoletus*: the role of specific stimulatory proteins, *Science*, 174, 293, 1971.
8. **Hasler, A. D. and Wisby, W. J.,** Use of fish for the olfactory assay of pollutants (phenols) in water, *Trans. Am. Fish. Soc.*, 79, 64, 1950.
9. **Hasler, A. D.,** Odor perception and orientation in fishes, *J. Fish. Res. Bd. Can.*, 11, 107, 1954.
10. **Hasler, A. D. and Wisby, W. J.,** Repelling Fish by Treatment with Potassium Phenylacetate, U.S. Patent 2,880,133, 1959.
11. **Idler, D. R., Fagerlund, U. H., and Mayoh, H.,** Olfactory perception in migrating salmon. I. L-Serine, a salmon repellent in mammalian skin, *J. Gen. Physiol.*, 39, 889, 1956.
12. **Little, E. E.,** Chemical communication in maternal behavior in crayfish, *Nature (London)*, 255, 400, 1975.
13. **McBride, J. R., Idler, D. R., Jonas, R. E. E., and Tomlinson, N.,** Olfactory perception in juvenile salmon. I. Observations on response of juvenile sockeye to extracts of food, *J. Fish. Res. Bd. Can.*, 19, 327, 1962.
14. **Muller, D. G. and Jaenicke, L.,** Fucoserraten, the female sex attractant of *Fucus serratus* L., *FEBS Lett.*, 30, 137, 1973.
15. **Pfeiffer, W.,** The fright reaction of fish, *Biol. Rev.*, 37, 495, 1962.
16. **Ryan, E. P.,** Pheromone: evidence in a decapod crustacean, *Science*, 151, 340, 1966.
17. **Teichmann, H.,** Concerning the power of the olfactory sense of the eel, *Anguilla anguilla* L., *Z. Vgl. Physiol.*, 42, 206, 1959.
18. **Todd, J. H., Atema, A., and Bardach, J. E.,** Chemical communication in social behavior of a fish, the yellow bullhead (*Ictalurus natalis*), *Science*, 158, 672, 1967.

Chapter 41

THE RECEPTORS

The more things change the more they remain the same.

Alphonse Karr

In Chapter 40 the olfactory receptor cells of land animals and insects were described briefly, and it is interesting to find much the same sort of structures in the receptors of fish. L. H. Bannister[1] made a detailed study of their fine structure in two species of teleostean (bony) fish, *Phoxinus* and *Gasterosteus,*

These cells are typical vertebrate olfactory receptors consisting of bipolar neurons with thin axons and thicker, but tapering dendrites, the tips of which are usually slightly swollen and form the receptor surface. Internally, all dendrites contain longitudinally oriented microtubules, or neurotubules. The presence of these within a cell process of dendritic dimensions is taken as an indication of that cell being a receptor. . .

The receptor cells fall into three categories, distinguished by the form of their distal tips . . .

The first type comprises the ciliated olfactory receptors each of which possesses a ring of 4 to 6 cilia upon its convex distal end. . .

The second type of receptor ending bears a variable number of microvilli instead of cilia. In other ways, however, it closely resembles the ciliated form and possesses a number of randomly oriented centrioles buried at various depths in the cytoplasm of the dendrite. The presence of these structures is strong evidence that this type of cell is an immature form of the ciliated receptor. . .

The third type of receptor ending I found only in *Phoxinus,* and appears to be unlike any olfactory structure that has been described before in vertebrates. This type bears neither cilia nor microvilli, but extends simply as a naked rod from the epithelial surface. . . Internally this type of receptor resembles the ciliated type in the presence of longitudinally arranged microtubules and mitochrondria, and of small vesicles.

Summing up, he concluded that,

In the present study it has been shown that in *Phoxinus* there are at least two types of olfactory receptor endings, one ciliated and the other rod-shaped. . . The majority of endings in the two species studied here, as found in a variety of other teleost species. . .conform to the pattern found in other groups of vertebrates.

The significance of there being more than one type of ending in one species is not clear. . .

Some of these differences, such as the gross form of the cell body and dendrite, appear to be rather superficial. . .

The receptor with a rod-shaped ending does, however, constitute a positive case of a specific chemoreceptor, in view of its peculiar internal structure. On the other hand, it may merely represent another method of increasing the area of the sensory surface. . .

An additional possibility is that the rod-shaped endings, since they project farther into the lumen of the olfactory organ, could receive chemical stimuli not affecting the ciliated receptors, especially when strong water currents depress the cilia of the latter into the mucus overlying the epithelium. Lastly, it may be argued that this type of cell is innervated by the trigeminal nerve, and that it constitutes a gustatory receptor.*

Generally similar results have been described by Breipohl et al.[3] once more demonstrating the essential similarity of the olfactory interface in the vertebrate series.

Naturally, in addition to these structures which are apparently involved directly in interactions with stimulus molecules, there are other structures needed to keep the system operating effectively. These include supporting cells, goblet cells which generate mucus, and most probably also a range of antibiotics to protect the bare nerve endings (with their direct pipelines to the brain) against the onslaughts of infectious viruses

* From Bannister, L. H., *Q. J. Microscope. Soc.*, 106, 333, 1965. With permission.

and bacteria. The need for a repair and maintenance service is all too readily overlooked by investigators whose attention is primarily directed towards the detailed structure of the receptors.

Moving up from the microscopic picture, the smelling organs of fish are, naturally, adapted to an underwater environment and differ appropriately from those of land animals. In the latter, the receptors are located in the nose where air is being drawn through continuously by the normal process of breathing and more especially by sniffing. In the fish, the receptor epithelium is normally situated in a pit on each side of the head, and has a lamellar structure — something like a book lying on its spine with its pages projecting loosely upward with spaces between for the water to flow through. The water currents may be generated in more than one way: in some cases there are two openings to the outside so that water enters through one and leaves through the other as the fish swims along. In others there may be motile cilia to sweep the water over the receptive surfaces. In still others there may be a pumping mechanism in the walls of the pit. These adaptations evidently have less to do with transcribing the message than with welcoming the messenger.

The fact that for the fish the messenger comes in a watery carriage makes it possible for a different sort of messenger to operate. In a land animal, the first qualification a molecular messenger must meet is volatility. A nonvolatile chemical can never be carried to the nose by currents of air no matter what its other attributes may be. For the fish, this limitation is replaced by an equivalent qualification — solubility.

It is not surprising, therefore, that many of the chemical messengers are substances such as amino acids which are quite nonvolatile in air and hence completely odorless to us. The amino acid, l-serine, was mentioned as eliciting the fright reaction in salmon making their upstream spawning run, and the minnow's alarm substance is, according to Pfeiffer and Lemke,[14] a pterin. The rainbow trout, *Salmo gairdneri,* responds to solutions of a variety of amino acids, and especially to glutamine, methionine, or leucine, and to the l-isomers more strongly than to the d-isomers. Threshold concentrations are of the order of 10^{-7} to 10^{-8} *M* and are comparable to the threshold concentrations of many airborne compounds. With solubility replacing volatility as an essential prerequisite to olfactory efficacy, it is by no means surprising to learn that the parasitic snail *Nassarius obsoletus,* depends upon a glycoprotein from the body fluids of the oyster to guide it.

The ability of salmon to recognize the home-stream odor in a particular tributary to a river has been shown most probably to depend upon a mixture of amino acids originating in the resident fish population of the stream rather than from such things as water weeds or chemicals leached from the soil. The species of char, *Salmo alpinish,* which makes its home in the Salangen River system in northern Norway is divided into a stationary and a migratory population. The migratory ones go to sea when 3 to 4 years old and spend about a month in the coastal waters. It has been suggested that they are able to recognize the correct branch of the river system by emanations from their stay-at-home relatives.

With other species of homing salmon, it seems probable that they are similarly imprinted, but not so much with the smell of their con-specifics as much as with the resident, nonmigratory population of other fish species. Experiments have shown that the permanent fish population tends to keep within surprisingly narrow territorial bounds.

In discussing the sequence of responses by which insects locate the source of an odor, it was pointed out that there is a "grammar" as well as a "vocabulary" in the insect language. The same is true of the olfactory language of fishes. If they are to locate the source of an attractive food odor, or move upstream to a spawning ground, the water must be moving. Lobsters move about on the sea floor and so are more like land animals than free-flying insects or free-swimming fish. Their responses to odors have been studied under three kinds of condition. They will walk upstream to the source of an odor carried by a water current. They do not orient if the odor is distributed uniformly in water that is stationary. If an odor trail is established while the water is moving past a point source, and if the water movement then ceases, they can follow the trail. To do this they have an advantage over both the dog and the insect in that their receptors are located on very long antennae so as to be sufficiently far apart for one side to be stimulated significantly more strongly than the other when they find themselves in the edge of the scent trail. Thus they can respond by turning toward the side of maximum stimulation, something that is not possible for an insect whose antennae may be no more than a few millimeters apart, or a dog whose nostrils are at most about a centimeter apart. The insect can respond at the edge of an odor trail by a random turn which has a 50% chance of keeping it in the odor trail. The fish with its larger brain and some amount of memory function can, perhaps, make a trial turn and then turn back if it proves unsatisfactory. Much of their behavior is, however, apparently an inherently programmed, nonrandom pattern within relatively fixed parameters.

Once again, therefore, we are brought back to the role of odors as messengers which evoke a response that is related less to the chemical or physical attributes of the stimulus than to the meaning the receiving organism assigns to the information the chemical signal brings to it. Once again, then, the story does not end with the identification of the messenger or the electron microscopy of the receptor organs.

REFERENCES

1. **Bannister, L. H.,** The fine structure of the olfactory surface of teleostean fishes, *Q. J. Microsc. Soc.,* 106, 333, 1965.
2. **Bertmar, G.,** Scanning electron microscopy of olfactory rosette in sea trout, *Z. Zellforsch. Mikrosk. Anat.,* 128, 336, 1972.
3. **Breipohl, W., Bijvank, G. J., and Zippel, H. P.,** Surface structure of the olfactory glands of the goldfish, (*Carassius auratus*). A scanning electron microscope study, *Z. Zellforsch. Mikrosk. Anat.,* 140, 567, 1973.
4. **Breipohl, W., Bijvank, G. J., and Zippel, H. P.,** Scanning electron microscopy of the olfactory receptor in the goldfish, *Carassius auratus, Z. Zellforsch. Mikrosk. Anat.,* 138, 439, 1973.
5. **Döving, K. B., Dubois-Dauphin, M., Holley, A., and Fourdan, F.,** Functional anatomy of the olfactory organ of fish and the ciliary mechanism of water transport, *Acta Zool. (Stockholm),* 58, 245, 1977.
6. **Fibak, J.,** A histochemical study of the excretory activity of supporting cells of the olfactory epithelium in fishes, *Folia Morphol. (Warsaw),* 4, 291, 1953.
7. **Gerking, S. D.,** Restricted movements of fish populations, *Biol. Rev.,* 34, 221, 1959.
8. **Gurin, S. and Carr, E. W.,** Chemoreception in *Nassarius obsoletus:* the role of specific stimulatory proteins, *Science,* 174, 293, 1971.
9. **Hara, T. J.,** Olfactory responses to amino acids in rainbow trout, *Salmo gairdneri, Comp. Biochem. Physiol.,* 44A, 407, 1973.
10. **Idler, D. R., Fagerlund, U. H., and Mayoh, H.,** Olfactory perception in migrating salmon. I. L-serine, a salmon repellent in mammalian skin, *J. Gen. Physiol.,* 39, 889, 1956.
11. **Kleerkoper, H.,** Some aspects of olfaction in fishes, with special reference to orientation, *Am. Zool.,* 7, 385, 1967.
12. **McLeese, D. W.,** Orientation of lobsters (*Homarus americanus*) to odor, *J. Fish. Res. Bd. Can.,* 30, 838, 1973.

13. **Nordeng, H.,** Is the local orientation of Anadromous fishes determined by pheromones? *Nature (London),* 233, 411, 1971.
14. **Pfeiffer, W. and Lemke, L.,** On the isolation and identification of the alarm substance from the skin of the European minnow, *Phoxinus phoxinus,* L., *J. Comp. Physiol.,* 82, 407, 1973.
15. **Solomon, D. J.,** Evidence for pheromone-induced homing by migrating Atlantic salmon, *Salmo salar,* L., *Nature (London),* 244, 231, 1973.

Chapter 42

FISH PSYCHOLOGY

In this world things are complicated and are decided by many factors. We should look at problems from different aspects, not just from one.

Chairman Mao

Psychological materialism is the doctrine that mental process is an attribute, mode, or effect of material process. Whether or not this doctrine is accepted in toto, it must be at least partly correct as long as mental processes can be affected by such material agents as chemical tranquillizers or stimulants, or by anything as simple as a distracting noise or a blow on the head. The behavioral changes brought about by these and other external agents are a proper study for psychologists, whether the subjects be people, bees, or fish. Moreover, many of the techniques worked out with man or the higher animals can be applied equally well to the simpler and possibly humbler species.

Thus, Todd et al.[4] used the method of training by reward and punishment to teach blinded yellow bullheads, *Ictalurus natalis,* to discriminate between the odors of other individual bullheads. The reward was a bit of food and a small electric shock was the punishment. By presenting either the one or the other in association with the training odor, it was possible to establish a conditioned reflex in an experimental fish. In a similar way, Walker and Hasler[5] trained bluntnose minnows, *Hyborhynchus notatus,* to discriminate between the odors of various species of aquatic plant. By training the fish to recognize one species and then observing the response to a second species, they were able to estimate the amount of resemblance between the odors of the two kinds of plant.

As mentioned earlier, the phenomenon known as imprinting produces a long-lasting behavioral response pattern to a given stimulus, and without the need for a reward and punishment training schedule, provided the imprinting is carried out at a certain critical stage in the individual's life span. Dizon et al.[2] imprinted young coho salmon by exposing them to 10^{-5} ppm of the chemical morpholine, and found that they showed a significant electrical response in the brain when exposed to solutions of morpholine 10 months later.

Evidently, then, fish have a memory function just as we have and which may well operate on much the same mechanisms as ours. It is well-known that in humans, there are two sorts of memory: long- and short-term. Long-term memory is called upon in such activities as spelling, or reciting poetry, or in carrying on the routine activities of living. Short-term memory is typically called upon to retain a telephone number from the time it is looked up in the directory until it has been dialed. Insects as potential test subjects for the study of the memory function will be discussed in Chapter 48 and it is sufficient to mention here that much the same considerations apply to fish, except that the experimental facilities, aquaria, etc. are likely to be more expensive.

One interesting approach was explored by Oshima et al.[3] who injected such memory-blocking chemicals as puromycin or cycloheximide into the crania of homing salmon and then used electrical recordings from the olfactory bulb to determine whether the fish were still able to recognize the water from their home stream. The treatment markedly interfered with their ability to discriminate between the presented and the stored patterns, but there was at least a partial recovery with the passage of time. They interpreted their results as implying that long-term memory depends upon continued metabolism of RNA and continued protein synthesis.

Table 1
BEHAVIOR UNITS AND SEQUENCES

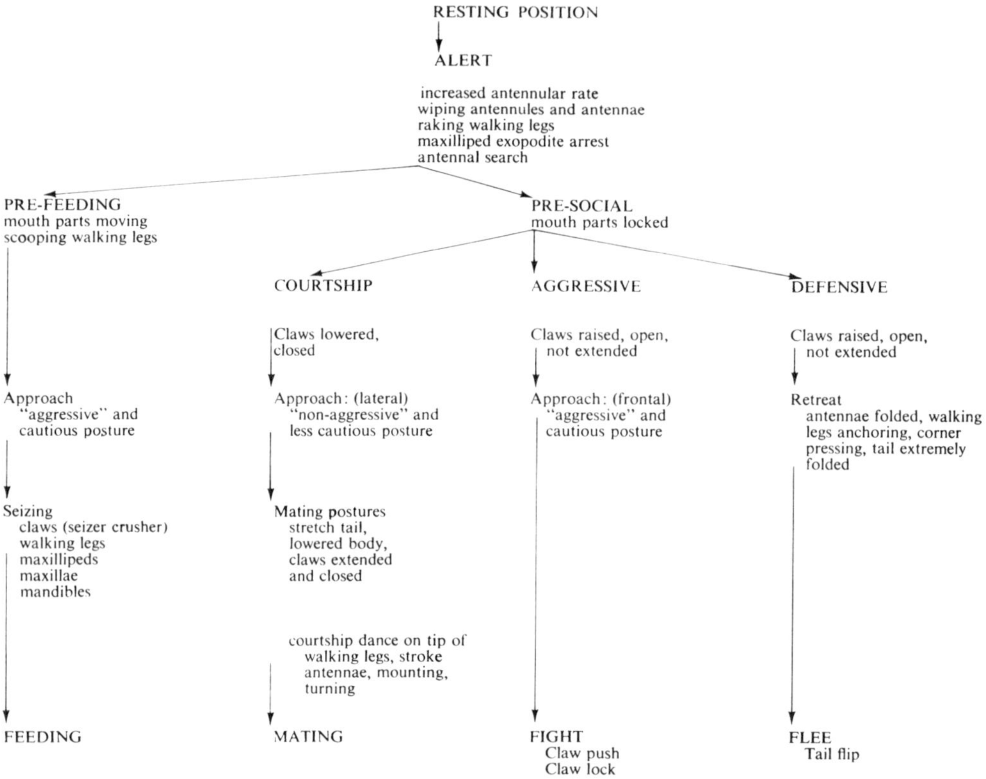

From Atena, A., and Engstrom, D. G., Reprinted by permission from *Nature (London)*, 232, 261, 1971.

This approach seems rather drastic and to involve what is perhaps a too-traumatic insult to the fish's bodily integrity. Work with insects in my laboratory convinces me that the most significant results were obtained when we observed insects which were completely unconstrained but which were exposed to very closely controlled sensory inputs. As was explained in Chapter 26, insect language has both a vocabulary and a grammar and a particular sort of behavior depends not only upon the nature of the sensory inputs but also upon their order of presentation. That similar response patterns can be seen in marine creatures is evident from Table 1 which is taken from a study of the lobster, *Homarus americanus*, by Atema and Engstrom.[6] It would be interesting to know whether the successive stages in the various kinds of response are programmed into the lobster's Central Nervous System (CNS) and come out in the right order in response to a single initiating trigger or whether several triggers have to operate in sequence. Either way, it seems evident that useful lights will be thrown upon the mental (or should one say the cerebral) process of a creature which is simpler than a rabbit but more complex than a honeybee.

At this point, I can hear an orthodox psychologist objecting that even if some of the neural and information-processing mechanisms of marine animals have interesting parallels with ours, there is little to be gained from diverting our attention away from the human aspects of the matter.

In saying this, he is overlooking one vitally important factor which can be summed up in one word: TEMPERATURE.

Any chemist knows that the key to understanding what is going on while chemicals are reacting, is to study the reaction rates *and the ways they vary with the prevailing temperature.*

Warm-blooded animals, including man, maintain their body temperatures constant within quite narrow limits so that it is a matter of great difficulty to find out whether the ability to recall an address or to retain an unfamiliar telephone number is in any way dependent upon the temperature of the brain. The problem is still formidable with cold-blooded reptiles whose bodily temperatures, while still less constant than ours, are nevertheless not alterable at (our) will. Insects while flying are necessarily metabolizing at a rather high rate and P. S. Callahan[1] has shown that the body temperature of one species of moth may rise nearly 9°F above ambient as a result.

Marine animals, whether they be vertebrate fish or invertebrate lobsters normally move slowly enough for their body temperature to be virtually identical with that of their watery environment, and they are usually tolerant to substantial changes in the water temperature. Thus they are, from a purely chemical standpoint, ideal subjects for an investigation of those activities of the brain that are likely to be comprehended by a chemist. To a geologist, the most valuable ores are usually found at the boundaries between different rock formations, and equally, the most valuable scientific insights are often to be found when one straddles the boundaries between different scientific disciplines. The chemical psychology or the psychological chemistry of fish is an almost unexplored boundary.

REFERENCES

1. **Callahan, P. S.,** A photographic analysis of moth flight behavior with special reference to the theory for electromagnetic radiation as an attractive force between the sexes and to host plants, Proc. 12th Int. Congr. Entomology, 302.
2. **Dizon, A. E., Horrall, R. M., and Hasler, A. D.,** Long-term olfactory "memory" in coho salmon, *Oncorrhynchus kisutch, Fish. Bull. Natl. Oceanic Atmos. Adm.,* 71(1), 315, 1973.
3. **Oshima, K., Gorbman, A., and Shimada, H.,** Memory blocking agents: effects on olfactory discrimination in homing salmon, *Science,* 165, 86, 1969.
4. **Todd, J. H., Atema, J., and Bardach, J. E.,** Chemical communication in social behavior of a fish, the yellow bullhead (*Ictalurus natalis*), *Science,* 158, 672, 1967.
5. **Walker, T. J. and Hasler, A. D.,** Detection and discrimination of odors of aquatic plants by the bluntnose minnow, (*Hyborhynchus notatus*), *Physiol. Zool.,* 22, 45, 1949.
6. **Atema, A. and Engtrom, D. G.,** Sex Pheromone in the Lobster *Homarus americanus, Nature (London),* 232, 261, 1971.

Chapter 43

BETWIXT AND BETWEEN

I take the liberty of recalling to my confrere, that the illusions of an experimenter form a great part of his power.

Louis Pasteur

Chapters 19 and 20 discussed the probable manner in which the vibrational attributes of a molecular stimulus generate inputs into selectively sensitive receptors so as to move a pattern of equivalent specificity into the Central Nervous System (CNS) for further processing. While it is doubtless premature to expect to comprehend the way (or ways) messages are transmitted from one part of the brain to another, it is not too soon to start looking into the matter.

Various experimenters have attached electrodes to the olfactory bulbs or other parts of the brains of salmon and other species of fish, and recorded the bursts of activity when variously scented streams of water were passed over the external olfactory epithelium. In general, there appears to be a conspicuous burst of electrical activity in some part of the CNS if the water carries something which is likely to convey a meaningful message to the fish. Ueda et al.[11] for example, reported that the strongest response was to water from the actual spawning site of the salmon they used, with weaker but clear responses to water from upstream or downstream from the site. Similar results were described by T. J. Hara[6,7] who used, in addition to home-stream water, solutions of various amino acids, including the mammalian skin factor, l-serine, which produces a sort of alarm reaction in migrating salmon.

These results leave no doubt that information is indeed making its way up a system of neural pathways from the olfactory epithelium to the control centers where the messages are interpreted and instructions sent out via the motor nerves. The experimenters are in many ways comparable to an external agency trying to monitor the activity of a telephone system by registering the total amount of traffic under various circumstances. In this way, the monitoring agency could show that the telephone traffic reaches a peak at certain times of the day, or seasons of the year, such as Christmas, or given such special circumstances as elections, tornadoes, Olympic Games, stockmarket crises, or Royal Weddings. The traffic patterns would vary in overall intensity and would usually be more or less localized in certain parts of the network, depending upon the nature of the crisis. Evidently this is an important phase in understanding the system as a whole and at this level of the investigation the analogy with olfaction is indeed very close. But it is still far removed from an understanding of the actual ways the individual messages make their way through the system, whether it be a system of wires or a system of nerves.

J. R. Hughes et al.[8] made an interesting attempt to take this further step with the olfactory message-transmissions. By implanting microelectrodes in the olfactory bulbs of rabbits and later monkeys he secured recordings which could be analyzed in various ways. The outcome was a claim that a specific stimulus generated a response which entered the glomeruli (Figure 2 Chapter 22) in the way suggested in Figure 1, Chapter 19 and was there transformed and sent onward via the mitral cells in a different way. The inputs from a specific type of primary receptor were fed into a specific mitral cell whose output consisted of a train of action potentials, evenly spaced at intervals which depended upon the type of primary cell where the signal originated. Eventually, he and his associates were able to make the same kind of experiments with human subjects.

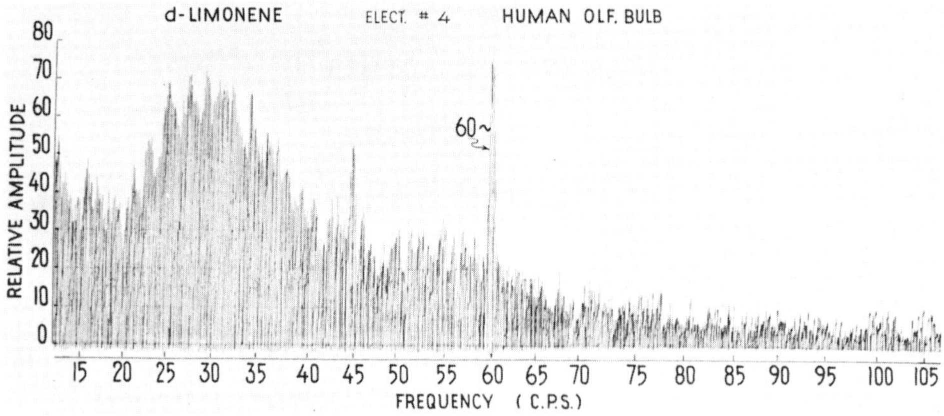

FIGURE 1. Analyzed response to an ororant (d-limonene). Note the separate peaks, especially between 20 to 50/sec representing different frequency components in patient H. M. Frequency is represented along the horizontal axis and relative amplitude along the vertical axis. The peak at 60 sec represents the presence of some 60 c line current in the recording, but these peaks helped to confirm the designation of the precise frequencies along the X-axis. (From Pfaffmann, Carl, Ed., *Olfaction and Taste,* The Rockefeller University Press, New York, 1969, 178. With permission.)

While animals can sometimes be taught or forced to communicate their sensations through their behavior, only man can describe his sensations in words. A rather unusual combination of circumstances made it possible for Hughes et al.[8] to look for correlations between specific chemical stimuli and (1) the electrophysiological responses in the olfactory bulbs of six men and five women, and, (2) the patients' verbal descriptions of the smell sensations they were perceiving. The experiments were evidently rather out of the ordinary and have been criticized on ethical grounds as having exploited human subjects. For the defense, it was pointed out that the patients were undergoing brain surgery for aneurisms or other conditions necessitating the exposure of the relevant parts of the neural network, the operating surgeons were in overall charge, and the patients fully understood what was intended and had agreed to cooperate. The electrodes could be left in place for several days after the operation, by which time the patients were conscious and able to tell the experimenters what the various stimuli smelled like at the same time that the electrical activity going on in their olfactory bulbs was being taped.

Subsequent processing of the taped records when different stimuli were being presented led to what was called the "Frequency Component Hypothesis" which suggests that the coding of odor quality at this level of the system takes the form of specific patterns or combinations of frequency components in the electrical output. Figure 1, taken from one of their reports, shows the analyzed response to a particular odorant, d-limonene. It suggests that the sensation in question is associated with a certain set of frequencies mostly falling into the range 20 to 50 Hz, or cycles per second. Other stimuli give different patterns, but with frequencies in the same general range.

In Chapter 20, the primary receptors were pictured as sending "there" or "not-there" inputs via parallel axons to the olfactory bulb where the inputs from primary receptors of a single type were collected and combined to be passed onward via one of the mitral cells. If Hughes' frequency component hypothesis is correct, each mitral cell should carry a signal made up of action potentials or spikes which should be equally spaced at intervals which are different for each type of primary cell and associated mitral cell. On this basis, it is difficult to see how the intensity is coded, but on the other hand, such a mechanism would do away with one of the problems raised by Hainer et al.[5] in their pioneering paper on information and olfaction.

This was the problem of reconciling the random arrival of molecules at the primary receptors with the presentation of a coherent signal to the CNS. The signal that arrives at the glomeruli (Figure 2 Chapter 22) has the form of irregularly spaced spike potentials coming from selected primary receptors. If the signal that moves onward via the mitral cells consists of trains of equally-spaced spikes, with the spacing dependent upon the frequency sensitivity of the primary receptors where that signal originates, then overall, the CNS is receiving wave trains of various frequencies which, together, make up a distinctive chord. If so, in the course of biological evolution, the neural processing of smell sensations may have become the model on which sound sensations are processed.

There has been surprisingly little follow-up to this approach to the problem, partly, I suspect, because it goes beyond conventionally conservative scientific thinking and partly because the conditions of the experiment inject so much background noise into the recordings that it is possible to dismiss the kind of results shown in Figure 1 as being merely random discharges.

However, Kaji et al.[9] have recently made similar spectral analyses of the olfactory responses of adult chum salmon to home-stream water and concluded that, "the results substantially agree with the frequency-component hypothesis proposed by Hughes. . .". Contrarily, K. B. Döving and R. Belghaug[3] have found that the frequencies derived by the Hughes method are shifted up or down scale by changes in the temperature of the olfactory bulb. His experiments were carried out with recordings from the surface of the olfactory bulb of the char, *Salmo alpinus*, L., and he concluded that the neural coding of the olfactory language is probably not reflected in the spectral patterns because, "the quality of the sensation would change with the temperature of the animals's environment."[3] An objection to this objection can be based on the musical analogy: the chord of C-major is recognizably similar to the chord of D-major (two semitones up the scale), and if a change in the brain's temperature speeded up our biological clock what we call C-major at one temperature would be identical with what we would call D-major at another.

It would be interesting to test a musician with absolute pitch when the body temperature was altered during a fever or while soaking in a hot bath.

It remains to ask whether the frequency components in the olfactory bulb can be correlated with the vibrational frequency patterns of the molecular stimuli. First, however, it must be made clear that the two kinds of frequency are both quantitatively and qualitatively different. The molecular vibrational frequencies shown in Table 1, Chapter 4 are of the order of 10^{13} to-and-fro movements per second, whereas the neural frequencies revealed by the analyses of the brain recordings are the result of action potentials passing along nerve fibers at intervals of more than about 0.01 sec.

In the hope of establishing a correlation between the frequency-patterns of the stimulus and the response-patterns inside the living organism, S. Yathom and F. E. Kellogg in my laboratory made recordings from the deuto-cerebrum of the desert locust, *Schistocerca gregaria,* when the antennae were stimulated by various perfume chemicals and also by the emanation from freshly cut grass. The neural responses were recorded on tape and analyzed in a manner similar to that used by Hughes, so that the results were not unlike those shown in Figure 1.

It was admittedly difficult to identify the frequency — components against the background noise, and more than one person who looked at the results dismissed them as simply random outputs. However, in the belief that in the dark even a (partly) imaginary light is better than no light at all, we provisionally identified a certain pattern in the response to grass and other patterns were seen (if only by the eye of faith) in the responses to some quite unrelated chemicals.

When fourth instar, wingless locust hoppers were exposed to the smell of grass in a wind tunnel, they responded by migrating to the upwind end. When they were presented with the various perfumes one at a time they made no such response. When a combination of perfumes were prepared which offered more than half the dozen or so components of the grass pattern, the hoppers once more began moving toward the upwind end of the tunnel. The artificial mixture was a blend of three compounds none of which is known to be present in grass, namely, *p*-methylacetophenone, musk xylol, and roseacetone.

It would be enchanting to think that we had intercepted signals passing into the insect's brain, decoded them, and then encoded a completely artificial signal that elicited the same sort of response, but some reservations were and are called for. While the response was unequivocal, the fact that we were simulating a rather general sort of food odor makes it possible that almost any mixture of chemicals might work in the same way. Unfortunately an unexpected illness by a member of the group interrupted the work and it was never resumed before I was forced to retire.

Some other pieces of unfinished business will be described in the following Part III of this book.

REFERENCES

1. **Dizon, A. E., Horrell, R. M., and Hasler, A. D.,** Olfactory electroencephalographic responses of homing coho salmon, *Oncorrhynchus kisutch,* to water conditioned by conspecifics, *U.S. Natl. Mar. Fish Serv. Fish. Bull.,* 71(3), 893, 1973.
2. **Döving, K. B. and Hyvärinen, J.,** Afferent and efferent influences on the activity patterns of single olfactory neurons, *Acta Physiol. Scand.,* 75, 111, 1969.
3. **Döving, K. B. and Belghaug, R.,** Frequency of induced waves in the olfactory bulb of char (*Salmo alpinus* L.); a temperature dependence, *Comp. Biochem. Physiol.,* 56A(4), 577, 1977.
4. **Döving, K. B. and Thommesen, G.,** Some Properties of the Fish Olfactory System, Olfaction and Taste VI., Paris, 1977, 175.
5. **Hainer, R. M., Emslie, A. G., and Jacobson, A.,** An information theory of olfaction, *Ann. N.Y. Acad. Sci.,* 58, Art. 2, 158, 1954.
6. **Hara, T. J.,** An electrophysiological basis for olfactory discrimination in homing salmon: a review, *J. Fish. Res. Bd. Can.,* 27(3), 565, 1970.
7. **Hara, T. J.,** Electrical responses of the olfactory bulb of Pacific salmon, *Oncorrhynchus nerka* and *Oncorrhynchus kisutch, J. Fish. Res. Bd. Can.,* 29, 1351, 1972.
8. **Hughes, J. R., Hendrix, D. E., Wetzel, N., and Johnston, J. W.,** Correlations Between Electrophysiological Activity from the Human Olfactory Bulb and the Subjective Response to Odoriferous Stimuli, Olfaction and Taste III, New York, 1969, 172.
9. **Kaji, S., Satou, M., Kudo, Y., Ueda, K., and Gorbman, A.,** Spectral analysis of olfactory responses of adult spawning chum salmon (*Oncorrhynchus keta*) to stream waters, *Comp. Biochem. Physiol.,* 51A(4), 711, 1975.
10. **Thommesen, G.,** The spatial distribution of odour induced potentials in the olfactory bulb of char and trout (Salmonidae), *Acta. Physiol. Scand.,* 102, 205, 1978.
11. **Ueda, K., Hara, T. J., and Gorbman, A.,** Electroencephalographic studies on olfactory discrimination in adult spawning salmon, *Comp. Biochem. Physiol.,* 21A, 133, 1967.
12. **Wright, R. H.,** Some alternatives to insecticides, *Pestic. Sci.,* 1, 24, 1970.

PART III
Possibilities

Whatever has come in industrial applications has come incidentally from experiments followed for their interest alone.

Irving Langmuir

Chapter 44

SULFURETTED HYDROGEN

There are some errors which, having been misunderstood for a long time, impair or retard the development of very large fields of science.

R. Taton

Two specific grounds on which the vibrational theory has been deemed heretical and consigned to the flames were considered in Chapters 13 to 15. There remains a third charge: one based on the fact that the lowest vibrational frequency in the Raman spectrum of hydrogen sulfide is at 1290 cm^{-1}. This is high enough to be outside the probable range of osmic frequencies, and so it is argued that the notorious rotten-egg smell of this compound cannot be attributed to its vibrational character.

It can be argued for the defense that hydrogen sulfide is chemically reactive and so the sensation we perceive may be engendered by the product of some reaction between it and a component of the olfactory apparatus. To account for its very low threshold, however, the reaction product would have to be a gas because anything else would operate only on a one molecule, one receptor, once basis.

It has many times been remarked that minute traces of impurities can modify the odor of a chemical sample, and so it seems worthwhile to explore the possibility that the smell we perceive is not in fact due to molecules of dihydrogen monosulfide, H_2S, but to contaminants regularly present in ordinary samples of hydrogen sulfide or formed en route to the olfactory receptor surface. I have a strong suspicion that these traces contaminants are simply homologs of H_2S.

Like other elements near carbon in the periodic table, sulfur atoms can link together to form chains, so that along with the paraffin hydrocarbons, the boranes and the silanes, there is an homologous series of sulfanes, H_2S, H_2S_2, H_2S_3, H_2S_4. These homologs are formed in several ways, including simple oxidation in the presence of air, moisture, and daylight. They have been studied in detail by F. Feher[2,3] and his colleagues, and fortunately their scrutiny included the Raman spectra of the individual members of the series. Table 1 shows the frequencies below about 750 cm^{-1} of the first eight members.

It is at once apparent that even though dihydrogen monosulfide has no vibrational frequencies low enough to evoke a smell sensation, the others could and, indeed, they are known to. Furthermore, the very low fundamental frequencies of the homologs higher than dihydrogen disulfide would suggest that their thresholds of perception would be very low also.

From Figure 1, which is based on Laffort's data for the olfactory thresholds expressed as P_{ol}s, the value for H_2S_2 would be about 9.0 and for the higher homologs they would be in the range 10.9 to 11.2. For hydrogen sulfide Laffort gives $P_{ol} = 9.72$. Taking these values at their face value, it would appear that less than 10% of any single homolog higher than H_2S_2 would account for the threshold of otherwise odorless H_2S, and if more than one were present their effects would be additive and the required level of impurity correspondingly reduced.

It is most desirable that ordinary hydrogen sulfide (to be better known hereafter by its old name, sulfuretted hydrogen) should be carefully analyzed, preferably by vapor-phase chromatography, and preferably using a split-stream instrument so that the odors of the fractions can be ascertained as they emerge from the column. It would be prudent also to test the fractions chemically. In a preliminary experiment I was able to carry out using less than ideal, borrowed apparatus and using "pure H_2S" from a

Table 1
RAMAN FREQUENCIES BELOW 750 cm^{-1} OF H_2S AND ITS HOMOLOGUES

H_2S	H_2S_2	H_2S_3	H_2S_4	H_2S_5	H_2S_6	H_2S_7	H_2S_8
None	509	483	483	485	484	464	484
		210	450	464	464	439	464
			229	439	439	242	439
			185	242	242	217	242
				217	217	185	217
				184	185	149	185
				149	149		149

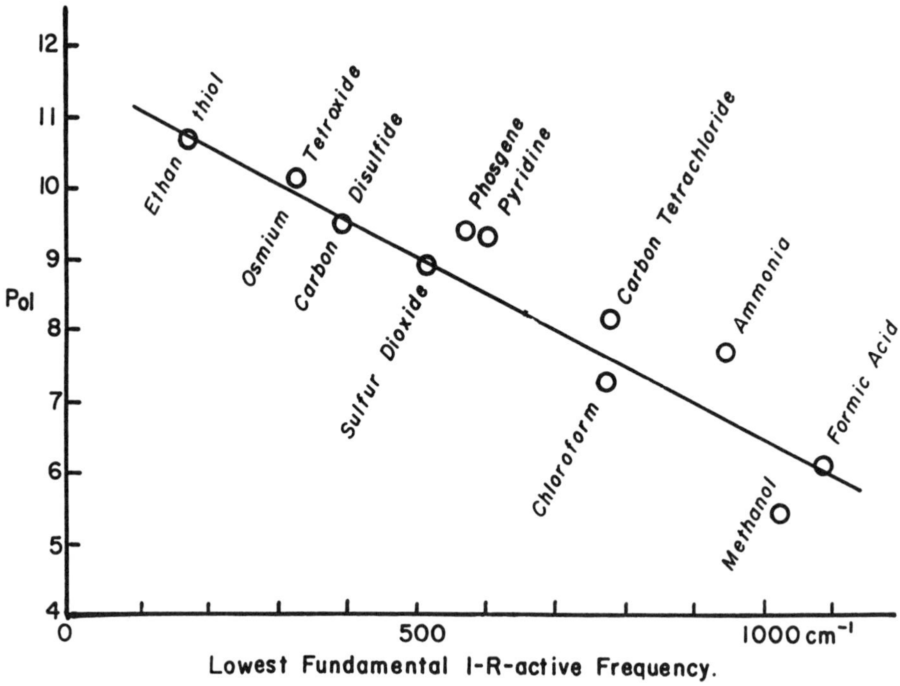

FIGURE 1. Laffort's threshold P_{ol} values plotted against available data for ν min.

cylinder, the largest peak had no odor but it also failed to blacken lead acetate paper so that there was at least one nonsulfurous contaminant in the supposedly pure hydrogen sulfide.

In a discourse on the misleading experiment, the late W. D. Bancroft[1] remarked that "sometimes the gods are good to the research man, though not often", because "all who do research know how difficult it is to make an experiment come out right when we know how it ought to go."

REFERENCES

1. **Bancroft, W. D.**,The misleading experiment, in The Methods of Research, *Rice Inst. Pam.*, 15(4), 224, 1928.
2. **Feher, F., Laue, W., and Winkhaus, G.**, The chemistry of sulfur. XXX. Preparation of the sulfanes, H_2S_2, H_2S_3, H_2S_4 and H_2S_5, *Z. Anorg. Allg. Chem.*, 288, 113, 1956.
3. **Feher, F. and Winkhaus, G.**, The chemistry of sulfur. XXXI. The preparation of the sulfanes, H_2S_5, H_2S_6, H_2S_7 and H_2S_8, *Z. Anorg. Allg. Chem.*, 288, 123, 1956.
4. **Laffort, P.**, Attempted standardization of olfactory thresholds of humans for 192 pure compounds, *Arch. Sci. Physiol.*, 17, 75, 1963.

Chapter 45

ARTIFICIAL NOSES

The electron has conquered physics, and many worship the new idol rather blindly.

H. Poincare

From time to time the claim is made that some new tool or instrument of the analytical chemist is a sort of artificial nose and may in some way parallel the nose's twin attributes of sensitivity and discrimination. Examples of such devices are the vapor-phase chromatograph and sundry ways of exploiting the dielectric properties of adsorbed films.

It seems clear that unless or until such a device is vectored to the special molecular attributes which confer olfactory specificity, it will be of more interest to the analyst than the osphresiologist.

Given that the relevant molecular attributes are the low-frequency molecular vibrational modes, it seems evident that for the moment, a genuine artificial nose must employ infrared radiation in one way or another.

C. W. Skarstrom[4] made an ingenious, nondispersive device which rivalled the nose in selectivity but not sensitivity. He put the gas mixture to be analyzed for a single component in a chamber where it could be alternately compressed and expanded so as to produce small fluctuations in temperature. The pulses of infrared radiation coming from the various molecular species in the mixture were all in the far infrared because the temperature never deviated much from the ambient value.

The pulsed radiation coming from the gas was divided into two beams which were directed onto the two ends of a thermopile. With no intervening filters the thermopile registered no signal. If the device was to analyze for a single component, say butane, in a complex mixture, an interference cell containing everything likely to be present *except* butane was interposed between the sample cell and both detectors, and a second cell containing butane was interposed at one end of the thermopile. The mismatch between the radiation reaching the two ends of the thermopile served to measure the amount of butane in the sample.

The device was extremely specific in its responses, but lacked the sensitivity needed in anything that could be in any way comparable to the nose. As an analytical tool, however, it was probably the first to exploit the genuinely osmic attributes of molecules.

With the development of infrared absorption spectrophotometers capable of working in the far infrared, it became possible to record and identify the vibrational attribute-patterns of odorous substances in the way described in Part I, but where mixtures are concerned, difficulties arise in filtering out the radiation-patterns of individual components, and the whole process is necessarily laborious and infinitely slow when compared with the almost instantaneous sensory input from a sniff. The ability of the brain to carry out an enormously complex pattern-analysis has been mentioned and is unlikely to be even approximately matched by any electronic gadgetry for some time to come. Therefore, I am inclined to think that for the time being the best approach will be to use the human brain to process the input from a detecting instrument. One possible approach might be as described below.

One class of far infrared spectrophotometer is based on the Michelson Interferometer and produces an interferogram in which the information as to the way the energy varies with the frequency is recorded as a function of the distance one mirror is moved with respect to another. The moving mirror is moved at a constant speed, so that the actual interferogram comes out as a plot of intensity vs. time which is also intensity vs. distance. Normally, the interferogram is fed into a computer which makes a Fourier Transform so that the result comes out as intensity vs. the reciprocal of the distance which is frequency. This gives the familiar type of absorption spectrum such as was shown in Figure 1, Chapter 5.

I believe it would be interesting to listen to the untransformed interferogram by recording it on magnetic tape and playing it back through a loud speaker, possibly with various tape speeds. I have no idea what it would sound like, but I do know I can listen to one instrument in an orchestra or one conversation at a cocktail party by focusing my attention appropriately. Therefore, it might be possible to pick out from the audible playback the contribution from one component of the mixture.

As I have more than once remarked, one thing leads to another. Not long ago I was telephoned by the producer of a television show who wanted me to do a 3 min interview in which I was supposed to answer the following questions about the problem of detecting terrorist bombs in airports and elsewhere:

1. Why use dogs?
2. How does their sensitivity compare with that of other animals?
3. Why use German Shepherds?
4. Are there any machines that could be used instead of dogs?
5. What are the relative advantages of machines vs. dogs?

For all this to be covered in 3 min would be, as Euclid long ago remarked, absurd. Therefore I decided to use my 3 min to say something that would be sensible and relevant to the subject.

The simpler insects have genetically-fixed responses to specific odors, usually manifested by a conspicuous clustering in the vicinity of the odor source, which may be of extremely low intensity. T. Kikuchi[2,3] isolated a mutant strain of fruit flies, *Drosophila melanogaster,* which was strongly attracted to at least 18 chemicals that are repellent to the natural population of wild fruit flies. Therefore, it seemed reasonable to suppose that a systematic search would produce a mutant strain which would be selectively attracted to certain explosives or various forms of contraband. If so, then all that would be needed would be an establishment to rear a continuous supply of the mutants and release a few thousand of them every day into the area to be monitored, and have everyone alerted to report suspicious assemblies of the insects.

Of course this is not an artificial nose, but given the existing state of the art it is a reasonable alternative possibility. For all I know, since the publication of the idea in *Nature,* work on it may already have been begun in some establishment concerned with security and antiterrorist measures. If so, we are unlikely to hear about it publicly for a long time after it is actually put to work.

REFERENCES

1. **Gebbie, H. A.,** Fourier transform versus grating spectroscopy, *Appl. Opt.,* 8, 501, 1969.
2. **Kikuchi, T.,** Specificity and molecular features of an insect attractant in a *Drosophila* mutant, *Nature (London),* 243, 36, 1973.
3. **Kikuchi, T.,** Genetic alteration of olfactory functions in *Drosophila melanogaster, Jpn. J. Genet.,* 48, 105, 1973.

4. **Skarstrom, C. W.,** Method and Apparatus for Determining a Component in a Gaseous Mixture, U.S. Patent, 2,794,924, 1957.
5. **Tanyolac, N., Ed.,** *Theories of Odors and Odor Measurement,* University of the Bosphorus, Bebek, Istanbul, (Contains an interesting description and discussion of various odor detectors.)
6. **Wright, R. H.,** Fruit flies as security officers, *Nature (London),* 260, 92, 1976.

Chapter 46

REFINING THE PATTERNS

Mathematics offers a wonderful shorthand for the precise formulation of well standardized ideas. On the other hand, the expressions of mathematics are lacking in humor, which is to say that they are no suitable medium for those finer shades of thought that are often necessary for the exposition of ideas which are on the way towards standardization.

G. N. Lewis

G. H. Cheesman devised a way to measure the community of odor property between two substances by observing how much prior exposure to one modified the perception of another that was sniffed immediately after. Two different techniques were used.

In one, rather low concentrations of both the adapting and the test stimulus were sniffed and the quantity measured was the amount by which the threshold of the test stimulus was raised. The apparatus needed to present the stimuli at precisely controlled concentrations was rather elaborate, and it required repeated tests by many people to produce a statistically valid result. In consequence, the cost/benefit ratio turned out to be rather disappointing.

An alternative and much simpler procedure was to fatigue the nose as completely as possible by several minutes exposure to a relatively high concentration of one substance, and immediately thereafter to sniff a second one. If the second one has normally an odor similar to that of the first, there will be a relatively large community of odor property and relatively little residual odor, and none at all if there is complete identity. If, however, the test stimulus includes additional notes or by-odors, they will be perceived rather clearly when the main component of the pattern has been cancelled out.

During a brief visit to my laboratory by Dr. Cheesman some years ago, we made a few experiments of the latter sort with a series of musk perfumes as adapting and test stimuli. With cyclohexadecanone as the adapting stimulus there was little or no residual with musk tibetene (1,2,6-trimethyl-2,5-dinitro-4-tert-butyl benzene) as the test, but when the roles were reversed cyclohexadecanone showed a definite, rather spicy residual. With cyclohexadecanone once again as the adapting stimulus and astratone (ethylene brassylate) as the test, there was a distinctly waxy or oily residual. When the roles of the pair were reversed, the musky odor of cyclohexadecanone was still perceptible, suggesting that it is a stronger musk than astratone. The waxy-oily component in the smell of astratone was verified when other musk perfumes were used as the adapting stimuli.

It would be interesting if a systematic study of this sort were to be made by someone with a trained odor memory and vocabulary who could therefore name specific compounds with odors resembling those of the residuals. An analysis of the various far infrared spectra might enable the patterns to be refined beyond the stage reached in Chapters 6 and 7.

As a variant of the foregoing, Figure 2 Chapter 6 seems to suggest that a frequency near 264 is somehow adverse to the perception of the musk odor. Before having to retire and leave my laboratory I was able to make a few experiments with the apparatus shown in Figure 1, Chapter 21 to investigate this point. The compound, *p*-methyl acetophenone (PMA) has a strong absorption maximum at 265 cm^{-1} and was tested against astratone. The musk odor appeared to be completely overwhelmed by a concentration of PMA of about 300×10^{-9} g mol/ℓ of air. At a concentration one tenth of this I could not detect the odor of PMA when tested by itself, but if I mixed it with astratone at this level and then adapted my nose to astratone I could smell the PMA in the mixture. If I adapted my nose to astratone plus PMA at this concentration in one channel, I could still smell pure astratone in the other, even though, without the PMA, adapting my nose in one channel prevented my perceiving it in the other. Altogether a very confusing picture, but one that would be well worth following up.

REFERENCES

1. **Cheesman, G. H. and Mayne, S.,** The influence of adaptation on absolute threshold measurements for olfactory stimuli, *Q. J. Exp. Psychol.*, 5, 22, 1953.
2. **Cheesman, G. H. and Kirkby, H. M.,** An air dilution olfactometer suitable for group threshold measurements, *Q. J. Exp. Psychol.*, 11, 115, 1959.
3. **Cheesman, G. H.,** Progress in odor research, Packaging Industry Research Association Report, The Research Association for the Paper and Board, Printing and Packaging Industries, Surrey, England, 1972.

Chapter 47

PRIMARY ODORS

The most fundamental concepts are the least definable, partly because of the inadequacy of language itself, but more particularly because we often wish to distinguish between things which differ in degree rather than in kind.

G. N. Lewis

In Chapter 28, a distinction was drawn between pure odors which are perceived by way of specialist receptors, and primary odors which are perceived by way of particular types of generalist receptors. Each kind of generalist contributes a primary element to the total sensation by being sensitive to a narrow range of frequencies calculated from the formula

$$F = 6.4N' \ cm^{-1}$$

where N′ is an odd integer.

In the ordinary course of events, a perceived sensation is never a primary one because even the purest chemical stimulus will present a pattern of vibrations rather than a single one, as indicated, for example, in Figure 3 Chapter 20. Nevertheless, it might be possible to generate a truly primary odor sensation in the following way.

It is known that olfactory stimuli presented at subthreshold concentrations can affect the smell of a mixture, presumably by a process of summation which alters the balance of primaries in the overall input. Of the compounds whose spectra are shown in Figure 3, Chapter 20, four, namely methyl phenylacetate, anethole, methyl salicylate, and ethyl benzoate, have a frequency close to the value 326.4 computed from the formula and postulated to be the sensitivity value of one type of generalist receptor. If the threshold concentration of each of these chemicals were determined, a mixture could then be prepared in which each was below its own threshold, but the common frequency would be above threshold and therefore perceptible as a primary odor. The apparatus required for the experiment would be like that shown in Figure 1 Chapter 21.

This is by way of example only. Given an adequate file of far infrared spectra, it should be possible to prepare other mixtures to generate other primary odor sensations and thereby open the way to a deeper insight into the mental processes involved in the discrimination and evaluation of olfactory sensations.

REFERENCES

1. **Amirov, R. Z.,** Effect of sub-threshold olfactory stimuli on the sensitivity of the sense of smell in normal and pathological conditions (transl.), *Biol. Eksperim. Bio. i Med.*, 47(5), 562, 1959.
2. **Keith, E. S. and Powers, J. J.,** Determination of flavor threshold levels and sub-threshold, additive, and concentration effects, *J. Food Sci.*, 33, 213, 1968.
3. **Neuhaus, W.,** The odor threshold of odor-mixtures for the dog and their relation to the thresholds for unmixed odors, *Z. Vgl. Physiol.*, 38, 238, 1956.
4. **Neuhaus, W.,** The ability of dogs to discriminate mixtures of odors, *Z. Vgl. Physiol.*, 39, 25, 1956.

Chapter 48

INSECT PSYCHOLOGY

The successful scrutiny of the investigator reveals itself in finding out that which phenomena, apparently very diverse, have in common, and the results are the more brilliant as the parallelised phenomena appear at the outset more diverse.

W. Nernst

In Chapter 25 it was explained that insect language includes words, which is to say, sensory inputs, and also something akin to grammar when the Central Nervous System (CNS) processes the sequence of signals coming in through the various channels so as to find out whether there is a coherent message. Only then can it decide whether a response is called for and what sort of response would be appropriate. In the higher vertebrates the amount of processing is naturally much greater and more complex than is possible in something as small as a mosquito or a fruit fly, but it is reasonable to suppose that the processing will be similar in its essential mechanisms. It is for this reason that insect psychology may someday yield unexpected rewards.

For example, in Denmark, H. Mourier[6] described a novelty effect when a new object in the form of a 15 × 15 cm black tile was placed on the floor of a cage containing a resident population of houseflies, *Musca domestica*. When it was first put in, the tile appeared to be highly attractive and received many visits from the flies. Then, from minute to minute, the number of visits diminished as the flies came to regard the tile as a familiar part of their environment, and in about 20 min it was visited no oftener than any other part of their domain.

If the tile was then removed for anything up to about 5 min and then put back there was no novelty effect, but the effect gradually reappeared as the time was extended, and if the tile was out of the cage for more than 45 min it once more became a new object when put back.

I observed exactly the same sequence of responses when an attractive scent was released through a localized area on the floor of a cage containing an established population of houseflies. The time rates were the same as those reported by Mourier.

Evidently we have a method of making quantitative measurements of the rates of learning and forgetting by houseflies. It would be interesting to know how these rates would vary if the prevailing temperature were controlled at various levels. Flies are reasonably active at temperatures as low as 20°C and as high as 38°C. If it could be shown that their mental processes are, or are not, temperature-dependent, it would afford some insight into the mental processes at work. Van't Hoff's Rule says that the rate of a chemical reaction is approximately doubled by a 10° rise in temperature. Does this apply to the memory function of houseflies?

The possibility of applying precisely controlled stimuli and of making quantitative measurements of the response, offers possible insights into central processing mechanisms which are as novel as they may be revealing. P. N. Daykin,[1] for example, showed that mosquitoes isolated from all external influences spend most of the time resting quietly on the walls of their cage, but now and then make short flights triggered by some event in their CNS which is generated by a strictly random process.

With the techniques now available for observing and recording unit responses, it should be possible to sort out various levels of central processing. Starting with the very smallest insects whose behavior must be almost wholly preprogrammed genetically, one could work up through houseflies whose heads are large enough to accomodate a limited memory function, and thence onward to the social insects whose programs are many times more complex but still genetically determined in the main.

The experiments of Floru et al.[3] on the influence of psychotropic substances such as the amphetamines, LSD, or alcohol, on the behavior of hornets, *Vespa orientalis,* F., showed that their effects are generally similar to those seen with other test animals. This is consistent with the postulate that living creatures do indeed share essentially the same basic neural mechanisms.

The use of social and other insects in biological, pharmacological, and psychological research is of more than purely academic interest. If, for example, a particular psychotropic drug can be assigned to a specific place in the neural program, a radioactively tagged material could be employed to locate the site of its action in the tiny brain of an insect more easily than in the enormously larger head of a monkey or mouse. It may be useful also to point out that there are not likely to be the same objections made to experiments and surgical operations on cockroaches or locusts as there would be to similar experiments using rabbits, rats, or dogs.

REFERENCES

1. **Daykin, P. N., Kellogg, F. E., and Wright, R. H.,** Host Finding and Repulsion of *Aedes aegypti, Can. Ent.,* 97, 239, 1965.
2. **Daykin, P. N. and Kellogg, F. E.,** A two-air-stream observation chamber for studying the responses of flying insects, *Can. Ent.,* 97, 264, 1965.
3. **Floru, L., Ishay, J., and Gitter, S.,** The influence of psychotropic substances on hornet behavior in colonies of *Vespa orientalis* F., *Psychopharmacologia (Berlin),* 14, 323, 1969.
4. **Kellogg, F. E. and Wright, R. H.,** The olfactory guidance of flying insects. III. A technique for observing and recording flight paths, *Can. Ent.,* 94, 486, 1962.
5. **Kellogg, F. E., Burton, D. J., and Wright, R. H.,** Measuring mosquito repellency, *Can. Ent.,* 100, 764, 1968.
6. **Mourier, H.,** The behavior of houseflies (*Musca domestica* L.) towards "new objects", *Vidensk., Medd., Dan. Naturhist. Foren. Khobenhavn,* 128, 213, 1965.
7. **Wood, P. W. and Wright, R. H.,** Some responses of flying *Aedes aegypti* to visual stimuli, *Can. Ent.,* 100, 504, 1968.
8. **Wright, R. H.,** Some quantitative aspects of insect attraction, *Can. Ent.,* 98, 1114, 1966.
9. **Wright, R. H.,** Housefly Memory, *Can. Ent.,* 106, 223, 1974.

Chapter 49

MIXED FARMING

All things have their emanations.

Empedocles

A square mile of ground planted with cabbages is a square mile heavily infested with cabbages, and nature will seek to restore her balance by moving in an expeditionary force of caterpillars. To combat them we apply an insecticide which may adversely affect honeybees, and the blame is placed on the maker of the pesticide when it should properly be shared by an economic system which calls for such intensive cultivation of cabbages. Left to herself, nature would have diluted the cabbages with hollyhocks, or lupins, or clover, which, if they did nothing else, would make it harder for a caterpillar to find a second cabbage after devouring the first. But there may be more to it than this.

In Chapter 7, the antiattractant effect of substances with a molecular vibration frequency adverse to the attraction of a specific insect was described. It seems more than probable that this phenomenon is operating most of the time, and perhaps all of the time, as part of the so-called Balance of Nature, but has so far gone unrecognized.

In the early 1960s, M. L. Bobb[1] at the Virginia Agricultural Experiment Station, studied a short-lived outbreak of an insect, the Virginia pine sawfly, *Neodiprion pratti pratti.* The outbreak began when there was a sudden and enormous increase in the ability of the females to attract the males. During 1960, when the outbreak was at its height, practically every female he trapped had mated and carried fertile eggs. Three years later, when the outbreak had faded away, fewer than 3% of the females had mated. The unusual peak in female allure was never explained, and the matter was apparently not followed up because the economic incentive to study it disappeared along with the sawflies.

It is entirely possible that some plant or fungus or other occupant of the ecosystem was less than usually abundant during the outbreak and so was not emanating the normal amount of the antiattractant which normally keeps the sawfly population down to an acceptable level. There is no proof of this, of course, if only for the reason that an entomologist investigating a plague of sawflies would be unlikely to notice that the number of mushrooms or dandelions was smaller than usual.

I have been told by amateur gardeners that planting marigolds near roses will keep down the number of aphids infesting the roses.

A systematic study of the vibrational spectra of substances which attract each insect species to its host plant so as to identify the adverse frequencies could be correlated with a similar study of the substances emanating from a variety of plants which does not attract the insects, so as to establish a rational intermingling of plant species for their mutual protection.

REFERENCES

1. **Bobb, M. L.**, Apparent loss of sex attractiveness by the female of the Virginia pine sawfly, *Neodiprion pratti pratti*,'' *J. Econ. Entomol.*, 57, 829, 1964.

Chapter 50

INTUITION OR ECM?

To find the truth is a matter of luck, the full value of which is only realized when we can prove that what we have found *is* true. Unfortunately, the certainty of our knowledge is at so low a level that all we can do is to follow along the lines of greatest probability.

J. J. Berzelius

There are people who say they can sense feelings of sadness or hostility or other moods and emotions in people they are with, no matter how carefully these feelings may be outwardly concealed. Sometimes these people walk into a house and immediately declare that it has a wonderfully happy atmosphere or perhaps that it has been the scene of some tremendous sorrow. Even when they turn out to be correct, orthodox scientists regularly dismiss this sort of thing by saying that the person claiming the insights has a too-active imagination, or is trying to attract attention, or is some kind of mental freak, and the subject is changed as soon as possible. It is a case of scientific curiosity being paralyzed by preconceptions about what is possible and what is not. In fairness to the scientists one thing that puts them off is the words the highly sensitive people are likely to use. In a house "haunted" by happiness or sorrow, they are likely to say that it has been the scene of a tremendous release of nervous energy. Scientifically, this is a meaningless statement, and so the matter is dismissed not because of what was done but because of what was said.

We must always remember that what *is* is natural.

Something that does not happen very often may be unusual but that is not to say it is abnormal or unnatural. The probability of getting 13 cards of the same suit in a bridge deal is precisely 1 in 39,688,347,497, or, in round numbers, once in every 40,000 million deals. If there are in North America six million people who play bridge fairly regularly, and if they play 100 evenings a year, and if 30 hands are dealt each evening, the total number of bridge deals in North America in 1 year is about 6,000,000 × 100 × 30 or 18,000,000,000. With a probability of a one-suit hand 1 in 40,000 million, such a hand is almost bound to turn up once every 2 years. In fact, the laws of probability guarantee that this can and must happen. It is unusual but not abnormal. What would be abnormal would be for a one-suit hand never to turn up at all.

The next thing to remember is that if something happens on the average only once in a million trials or once in a million years, it may not be necessary to make a million trials or to wait a million years for it to happen. It is just as likely to turn up near the beginning as near the end or somewhere in between.

Because a thing is unusual it is not therefore abnormal or unnatural. Therefore it is a mistake to label such things as "paranormal" and to invent schools of "parapsychology" or "para-" anything else to deal with them. To do so usually leads to phrases being coined like "discharges of nervous energy" that are without meaning to orthodox scientists with the result that they quite naturally and properly relegate the parascientists to the lunatic fringe.

Which brings us back to the people who claim to be and may in fact actually be sensitive to the atmosphere of a person or a building.

It is likely that their use of the word "atmosphere" is not accidental and may, indeed, give us a clue to what is really happening.

In Chapter 30 we discussed the kinds of odorous emanations that might come from the bodies of persons in stressful situations, and in Chapter 31 it was shown that the personal scent can be transferred to the surface of various things. In Chapter 27 the possibility that certain people may really be exceptionally sensitive in their perception of such External Chemical Messengers (ECM) was presented.

In the light of these considerations, it would be legitimate to replace such phrases as a massive release of nervous energy with the words a massive discharge of ECMs. We would then be speaking in the language of orthodox science, and what is much more important, we would be stating the problem in terms that open the way to orthodox scientific investigation.

It is known that fright causes adrenalin to be released into the blood stream, and at least one study has shown that there is a distinct biochemical effect on the composition of the blood and urine of men, but not of women, while watching a pornographic motion picture.

Given that the existence of ECMs is a tenable hypothesis, the next step is to contrive experiments to test it. It might be done like this.

We would start with two groups of people, about a dozen in each group. The first group would be taken into a rather small room with rather poor ventilation and shown a 5 min film of a kind likely to arouse a powerful emotion of some particular kind. It could be something to make them laugh, or to make them angry, or to arouse their pity, or to excite them sexually, or maybe just to bore them profoundly. But it would have to be something likely to evoke a strongly felt mood in most of them. At the end of the film they would be moved out of the room as quickly as possible by one door, and immediately afterward the second group would be brought in through a second door and asked to guess what kind of mood the first group had experienced or what kind of film they had seen.

It is likely that most people would turn out to be making random guesses, but by retesting those who got the right answer more often than not, we might find a certain number of genuinely perceptive people who really can "sense the atmosphere" of a room and associate it with a certain kind of mood.

If this phase of the experiment works out, then an obvious next step would be to do careful chemical analyses of the atmosphere, and to reconsider some earlier opinions and judgments of persons and "facts".

At the very least, it would be fun.

Chapter 51

PIGMENT OR FIGMENT?

A scientific author must always envy the narrator who is not obliged to coerce into some artificial order the natural sequence of the story. To arrange in serial order a great mass of branching and interlocking ideas is always a painful task.

G. N. Lewis

Viewed with the naked eye, the exterior surface of the olfactory receptor organ stands out from the surrounding tissue by being colored. The color ranges from yellow through dark brown to nearly black. Figure 1 shows the olfactory end-organ of the Pacific salmon laid bare by cutting away the overlying fold of skin. From time to time the presence of this pigmentation has been made the point of departure for constructing a theory of the primary interaction process. One of the more recent ones was developed by Drs. Reid and Evans[15] with some assistance from me, but has by now sunk without a trace.*

The literature reports several attempts to identify the pigment. Experimentally, the problem is difficult for several reasons. For one, the physical structure of the olfactory epithelium with its sensory cells and cilia, supporting cells, goblet cells, blood vessels, and so forth, ensures the presence of a host of substances which have nothing to do *directly* with the actual triggering process. For another, because of its constant exposure to the external world, the olfactory membrane is constantly bombarded with microorganisms and other hostile entities so that degenerative changes set in exceptionally rapidly when life with all its protective activities ceases. It is therefore not surprising that there have been conflicting accounts of the kinds of substances contributing to the coloration.

Vitamin A or carotenoids related to it have been reported by some investigators but not by others using different assay methods. There are, however, substantial reports that the sense of smell is impaired in persons or animals suffering from a vitamin A deficiency. B. Rosenberg and his colleagues[14] suggested a transduction mechanism based on the fact that carotenoids are semiconductors whose semiconducting properties can be changed if odorous molecules become weakly bound to them.

Colored substances other than carotenoids have also been reported, including lipofuscins, phospholipids, and melanin, so that the field is still wide open. Some hitherto unpublished experiments I made with the olfactory tissues of Pacific salmon showed that the greater part of the coloration is due to quite large, irregularly shaped and dark colored objects imbedded in the tissues. They resembled melanophores and were not affected by solvents or even by solutions of caustic soda or sulfuric acid, and as far as I could determine they were simply accumulations of melanin. This is a rather inert, polymeric substance, which is responsible for the transient coloration of suntanned Caucasian skin or the permanently colored African skin, where its function is to protect the underlying tissue from excessive exposure to ultraviolet radiation. This has occasionally led me to wonder whether the pigmentation may be there to shield the receptors from photo-effects which could in some way impede the process of stimulation.

At least one writer has suggested that the pigment is nothing more than a waste

* Scientists are also human. When certain critics of the vibrational theory declared categorically that, "there is no conceivable way a low-frequency, low-energy vibration could trigger the discharge of a nerve cell", it became a matter of duty to conceive of one!

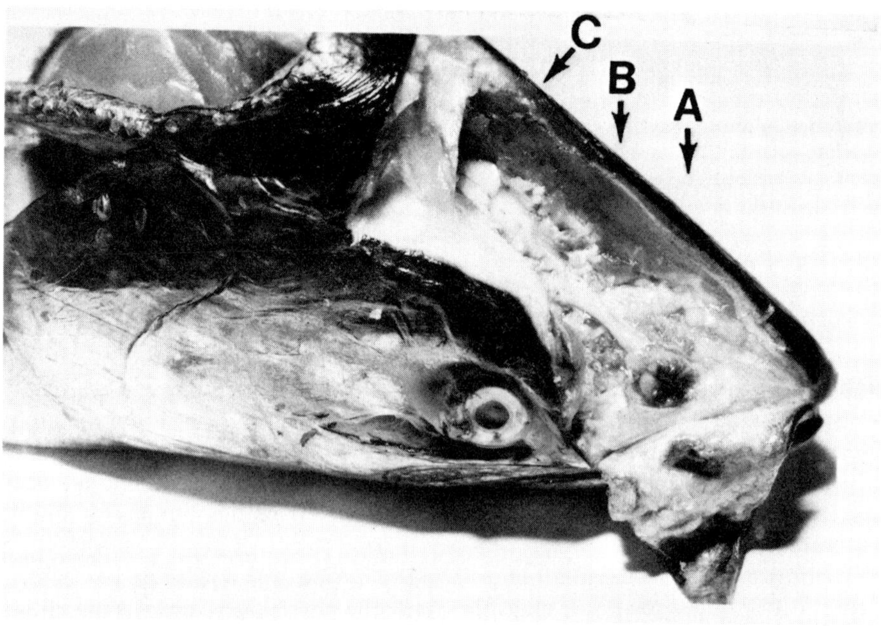

FIGURE 1. Dissection of a salmon head: The dark-colored receptors at A are connected by the olfactory tract, B, to the brain, C. Water from the outside flows over the receptor surface, entering by one opening and leaving by another.

product of the metabolism of lipids peculiar to the olfactory epithelium. As against this (or possibly connected with it) is the fact that some structures deep in the Central Nervous System (CNS) and forever hidden from the light of day are also seen to be dark colored when the skull is opened. In general, when a substance is viewed with visible light and seen to be colored, this is because it has electronic levels easily excited either metabolically or by quanta of visible light. It is by no means impossible that some of the processes going on in the CNS involve some kind of electronic excitation, and something of the sort may be happening in that part of the CNS which is exposed to the outside world (Chapter 22).

It has been claimed and also denied that olfactory acuity is correlated with the depth of the pigmentation. Claims that albinos have no sense of smell have been pretty well disproved, but this is not to say that their sensitivity is as good as that of normally pigmented animals or persons. K. Onegawa[12] used electrical recordings from the olfactory bulb of the rabbit to show that the response is higher when the color is dark. The need for some sort of shielding against daylight would be greater the nearer the receptor surface is to the outside. In the salmon, the receptors are located in shallow pits with two openings to permit water to flow through and over the receptors. Most terrestrial, air-breathing animals have the receptor surface situated some distance from the open nostrils and therefore more or less adequately shielded from the light. It would be interesting to determine whether olfactory acuity is reduced by illuminating the receptor surface in some way that does not injure the organism or impede the flow of air. In insects, where the receptors are located on the antennae, they are usually situated at the bottoms of pits in the dark-colored chitinous material of the antennae, and the most striking manifestations of insect olfaction take place during the night.

Evidently, the nature and role of the olfactory pigment present some fascinating problems.

Finally, the pigment is only one of the chemical components of the receptor apparatus. There are proteins, lipids, and a host of substances whose presence is in one way or another essential to the proper functioning of the mechanism. Besides the basic structural materials there must be enzymes to control a whole host of processes necessary for the repair and maintenance of a going concern, and, what seems to have been rather generally overlooked, a battery of antibiotics and other agents to protect this gateway to the brain from an army of hostile invaders.

It would be fatally easy to focus attention on one or the other of these components of the tissues, especially if it happened to be one for which there was a readily available bioassay or chemical test, and to consider it *the* clue to the process of stimulation.

A preconception used as a guide in making observations or experiments is called a working hypothesis and is one of the most useful tools of the scientist, but care is needed lest the preconception become an obsession. On balance, the history of science demonstrates that the magpie collection of facts without any expectations to guide the search is less productive than researches intended to test first one point and then another along a broad interface with the unknown.

REFERENCES

1. **Dodd, G. and Cash, C.,** Lipofuscins, ascorbic acid and the olfactory pigment, 1st ECRO Congr., Paris, July 3 to 5, 1974.
2. **Menevse, A., Dodd, G., and Poynder, T. M.,** Evidence for the specific involvement of cyclic AMP in the olfactory transduction mechanism, *Biochem. Biophys. Res. Commun.*, 77(2), 671, 1977.
3. **Duncan, R. B. and Briggs, M. H.,** Treatment of uncomplicated anosmia by Vitamin A, *Arch. Otolaryngol.*, 75, 116, 1962.
4. **Jackson, R.,** The olfactory pigment, *J. Cellular Comp. Physiol.*, 55, 143, 1960.
5. **Koch, R. B. and Norring, N. L.,** Fractionation of olfactory tissue homogenates. Isolation of a concentrated plasma membrane fraction, *J. Neurochem.*, 16, 145, 1969.
6. **Koyama, N., Sawada, K., and Kurihara, K.,** Isolation and some properties of plasma membranes from bovine olfactory epithelium, *Biochem. Biophys. Acta*, 241(1), 42, 1971.
7. **LeMagnen, J. and Rapaport, A.,** The role of Vitamin A in the mechanism of olfaction in the white rat, *C. R. Soc. Biol.*, 145, 800, 1951.
8. **Margolis, F. L.,** Carnosine in the primary olfactory pathway, *Science*, 184, 909, 1974.
9. **Menevse, A., Chang, L., Menco, B., and Dodd, G.,** Adenyl cyclase activities of olfactory and gustatory plasma membranes, 1st ECRO Congr., Paris, July 3 to 5, 1974.
10. **Moulton, D. G.,** The olfactory pigment, in *Handbook of Sensory Physiology*, Vol. 4 (Part 1), Springer-Verlag, 1971, 59.
11. **Nadol, J. B. and DeLorenzo, A. J. D.,** Fine structural localization of acetyl cholinesterase in the olfactory mucosa in guinea pig and rabbit, *Ann. Otorhinolaryngol.*, 83, 55, 1974.
12. **Onegawa, K.,** On the relationship between the action potential of the olfactory mucous membrane and the grade of the color of the olfactory mucous membrane, *J. Physiol. Soc. Jpn.*, 19(3), 194, 1957.
13. **Phillipot, E. and Gerebtzoff, M. A.,** First results of the analysis of the olfactive pigment, *J. Physiol. (Paris)*, 50, 451, 1958.
14. **Rosenberg, B., Misra, T. N., and Switzer, R.,** Mechanism of olfactory transduction, *Nature (London)*, 217, 423, 1968.
15. **Wright, R. H., Reid, C., and Evans, H. G. V.,** Odor and molecular vibration. III. A new theory of olfactory stimulation, *Chem. Ind.*, 1956, 37, 973.

Chapter 52

DÉJÀ VU

On hearing these words, Copperfield suddenly experienced the illusory recognition, which he described with an almost scientific exactitude as follows: We have all some experience of a feeling, that comes over us occasionally, of what we are saying and doing having been said and done before, in a remote time . . .

Charles Dickens

The phenomenon known as *déjà vu*, of feeling that we know perfectly what will be said or done next, as if we suddenly remembered it, is as well recognized as it is inexplicable. The feeling that something happening *now* is a total reenactment of the identical something experienced at some moment in the past is by no means a rare experience and the name *déjà vu* — meaning "already seen" — is intended to express the feeling that the individual is reliving a situation so that he knows what has happened as well as what is about to happen. The experience seldom lasts more than a minute or two, it cannot be evoked at will, and so it has fallen outside the compass of conventional psychology.

As a consequence, it has been seized upon by parapsychologists and others even farther out from the center of orthodox science as a manifestation of precognition, or second sight, or as evidence of a previous incarnation. The best that orthodox psychologists seem to have made of the phenomenon is to suppose that it represents a momentary dislocation in the time-sense so that what is being registered in the mind is alternated between the present and the past tenses. This seems to smack more of a description than an explanation of what is going on. An alternative but still an ad hoc explanation regards the phenomenon as an unexplained relationship between recognizing and recalling.

It is curious that the phenomenon does not seem to have been considered in relation to a rather similar and much better known phenomenon, namely, the "evocative" power of a smell sensation. This has been remarked upon many times: the perception of a distinctive odor will bring back a vivid remembrance of some previous event or occasion in which the smell played an important part. This happens when the odor was and is present at a high enough level to intrude upon the consciousness distinctly enough to be a part of the situation.

But suppose that the odor has been present at a rather lower level of intensity. It has been part of the occasion but not in a way that makes it obtrusive. Under those circumstances, its ability to evoke a recollection (literally, to reassemble the incidents) might operate in such a way that the incidents come up to the level of consciousness even though the odor itself lies unnoticed in the background.

Such an hypothesis seems essentially reasonable, given the recognized evocativeness of odors together with our socially conditioned disposition to relegate them to a secondary or even a tertiary place in our thinking.

It is reasonable to suggest that the matter could be approached experimentally in some such way as the following. The person chosen as the "subject" (that is, the "guinea pig") would be shown a film clip of a bit of unfamiliar scenery or allowed to overhear or take part in a snatch of conversation, all in accord with a scenario carefully planned, though not by him. At the same time, he would be exposed to a rather unusual, distinctive, but not an obtrusive odor. The occasion would have to be one that would not otherwise be especially or intrinsically memorable. At some future time, the subject would be "accidentally" exposed to a somewhat similar scenario simultaneously with the identical odor, and afterwards asked whether there was any sense of *déjà vu* in the experience.

If, for example, the original scene was a particular and fairly distinctive bit of country, or a rather special street scene, the reenactment might be a mention of such a visual imprint or the description of some incident in which the scene was an important factor. Whether or not the associated odor should be associated directly with the scenario will have to be determined by trial. If, for example, the scenario has an oriental flavor, should the accompaniment be one with an oriental association also, or should it be something incongruous and unrelated such as a foxy or sea-weedy sort of smell?

Even if the prospect of a positive, meaningful outcome is no better than 1 in 20, the experiment is worth trying because a positive result would help to resolve a present mystery and thereby illuminate a still-obscure mental phenomenon. A thing does not become less wonderful by becoming less mysterious. On the contrary, by resolving a small mystery we enlarge the greatest mystery of all — the fact that the Universe is comprehensible!

APPENDIXES

A text-book should be read forwards *and* backwards. J. W. Mellor

. . . the whole subject of smells and smelling is in a very healthy condition because there are more questions than answers, and more ways of looking for the answers than ever before. The next years should see new advances and interesting discoveries that will prove some of the "facts" and many of the ideas in this book to be wrong.[7] R. H. Wright, *The Science of Smell*

Appendix A

ANTECEDENTS AND ALTERNATIVES

Since the object of this book has been to present a coherent account of smells and smelling predicated upon the quantized, mechanical oscillations of molecules, it seemed best to plunge directly into the matter in Chapter 1, rather than to open with a historical recapitulation, perhaps starting as early as Albrecht von Haller's *Elementa Physiologiae* of 1763 and making passing reference to everything contributed subsequently. To do so would be an exhausting exercise at best and would almost certainly confuse the real issue which is simply this: is it at last possible to present a unified and comprehensive account of smell phenomena in all their ramifications?

At the same time, it is fair to say that not everyone is today in agreement with the theory presented here. R. W. Moncrieff in his book,[2] *The Chemical Senses* (2nd. ed., 1951) reviewed some 30 distinct theories, mainly from the point of view of a chemist. At about the same time, Jones and Jones in their Modern Theories of Olfaction: A Critical Review, (*Journal of Psychology,* 36, 207, 1953) covered much the same ground from the standpoint of psychology.

As it turned out, these reviews and commentaries were written just at a turning point in the development of the subject. In the quarter of a century since they appeared, the field has narrowed down to three currently competing theories, namely, the vibrational theory presented here, the stereochemical theory set forth in its most modern form by J. E. Amoore in his book *Molecular Basis of Odor* (Charles C Thomas, Springfield, Ill., 1970) and the structural theory of M. G. J. Beets presented in *Structure-Activity Relationships in Human Chemoreception* (Applied Science Publishers, London, 1978).

It would be invidious for someone committed to a rival theory to recapitulate their ideas and this is hardly the place to enter upon a critical commentary. It is for the uncommitted third party to compare each advocate's presentation and arrive at his own verdict.

At the same time, it is fair to say that not everyone is today in agreement with the theory presented here. R. W. Moncrieff in his book,[2] *The Chemical Senses* (2nd ed., 1951) reviewed some 30 distinct theories, mainly from the point of view of a chemist. the earlier writers. By disregarding entirely the olfactory responses of insects and making at most only passing reference to those of animals, they have cut themselves off from an enormous range and variety of experiments which are as relevant as they are significant. One is tempted to ask how much would be known of human genetics if Mendel's work with peas, and the myriads of studies on fruit flies were dismissed as irrelevant to the phenomena of human heredity?

The foregoing chapters will have shown that no branch of science (with the possible exception of astrophysics) can be left out of consideration in relation to some aspect or other of osphresiology. Even so seemingly remote a subject as mining engineering has been involved when trained dogs are used in Finland and elsewhere to prospect for valuable minerals.

The reader who has made a critical appraisal of current theories and wishes to work back into the past will find two valuable compilations of the literature prior to about 1960. *Odors and the Sense of Smell. A Bibliography, 320 B.C. to 1947* published by Airkem, New York, 1950, lists publications by author and title. Odor and Olfaction: A Bibliography, 1948—1960 by K. M. Michels et al. and published in *Perception and Motor Skills,* 15, 475, 1962, extends the Airkem list by another dozen years. There does not seem to have been anything comparable during the 1960s, but beginning in 1973, *Chemoreception Titles* and *Chemoreception Abstracts* published by Information Retrieval, Ltd. (London) help one to keep abreast of current developments. The proceedings of various symposia provide an overview of the matters that have occupied the attention of a variety of investigators, but on the whole do little to bring about a synthesis of their ideas.

What has happened in the past 25 years is a widespread interest concentrated upon one or another specific aspect of the subject. Earlier work has been reexamined and reevaluated, and inevitably found to contain much that is ephemeral but with more than a few real gems embedded therein. It is, of course, only on the basis of hindsight that the gems are recognizable, and it is only with hindsight also that we see that the gems could not have come into existence without the accompanying magma out of which they have crystallized.

REFERENCES

1. **Wright, R. H.**, *The Science of Smell,* George Allen and Unwin, Ltd., London, 1964, 158.
2. **Moncrieff, R. W.**, *The Chemical Senses,* 2nd ed., John Wiley & Sons, New York, 1951.

Appendix B

PROTOCOL OF A FOUNDATION FOR THE ENCOURAGEMENT OF RESEARCH IN OSPHRESIOLOGY

It is worthwhile to examine some ways a capital sum could be used to expand research in a given field, which, for concreteness will be taken to be the matter of smells and smelling.

At least three approaches can be considered.

An institute — This includes a building, an administration, library, workshops, and other services as backups for the people actually doing the research. Given a Director of the calibre of Clerk Maxwell or Ernest Rutherford the Institute might achieve the status of the Cavendish Laboratory, but the probability is small and taken together with the overhead costs the endowment of an Institute is probably not the best use for a limited capital sum.

An eleeymosynary foundation —Its function is to provide money for people in universities, etc. where the necessary backup facilities are already provided. The running expenses are less than those of an institute and so a larger proportion of the available money is available to the people actually doing the research.

The disadvantage of this approach is in the nature of advisory committees. Committee members feel some responsibility for making sure that money is not spent on frivolous projects and so they tend to keep well inside the bounds of conventional thinking. This in turn tends to result in torrents of *data* but is less successful in generating *ideas* which if they are new are necessarily unorthodox. It is fatally easy for a committee to make self-fulfilling predictions: if it decides that something won't work and it is never tried then naturally it doesn't work.

A novel approach — In most graduate schools there is at least one man who has an encyclopaedic knowledge of what everyone is doing in his field, and spends so much time collecting references and talking about it all that he does hardly any original work himself. The faculty regard his prospects as a researcher as rather poor and he is usually eased out of the school. I suggest that properly appreciated and used he might become an invaluable member of the scientific community.

Pay this man a salary, give him a travelling allowance, and charge him with the duty. of visiting universities and research institutes over a nationwide (or larger) area. His job is to learn who is doing what so that he can tell researcher A that researcher B on the other side of the continent has ideas that indirectly link up with those of A, and that C has a piece of apparatus that is not being used full-time and might become available to meet part of A's needs without A having to buy an expensive item.

I can illustrate both these functions by specific examples.

An infrared spectroscopist in Alberta read about my work on the vibrational theory of smell and wrote to ask whether he could help by running a few spectra for me. In reply, I put him in touch with an entomologist in Ontario who was interested in the vibrational patterns of some insect-attracting chemicals . . . A second example is incomplete: it concerns a university department that used some unexpended funds to buy an electron microscope which thereafter sat idle because there was no one to use it!

Most people doing research are bombarded with paper in the form of abstracts, journals, symposia, book notices and assorted advertisements, most of which go unread. A couple of hours or half a day spent with a good listener who may be able to suggest useful contacts outside one's own circle is a different matter. It will be many times more significant if the visitor has access to money with which to develop inter-institutional contacts and cooperation. Thus, for example, if there had been money to provide the infrared spectroscopist mentioned above with a graduate student or other assistant, a continuing program of collaboration would have replaced what was probably at most a one- or two-shot spare time affair. Or again, if there had been money available, the unused electron microscope might have been put to work for people in other departments or fields and its research potential realized before the passage of time rendered it obsolescent.

Given this method of promoting research, it is evident that the travelling director must have the authority to approve a grant without having to refer it to a committee. Naturally, he should have a properly prepared proposal and application in the usual way, except that he might have had a hand in drafting it.

Evidently, too, his terms of appointment should make it clear that he has a two-fold function. He has to encourage cooperation between specialists as in the case of the infrared spectroscopist and the entomologist. This is obvious. A second and possibly less obvious function is to attract new people into the field so as to open up new avenues. This would develop towards the end of his term of appointment as his understanding of its ramifications grows and as he has picked up casual hints from offhand remarks by those he visits.

As an example of the sort of thing that might eventuate, consider the following set of (apparently) unrelated items.

1. Workers in Germany identified two types of odor receptor on the antennae of insects. Generalists help the insect to manage its day-to-day activities such as feeding or nesting. Specialists respond selectively to species-specific sex-attractants and other especially meaningful signals.
2. A pediatrician in New York suggested that *some* cases of schizophrenia might be the result of an individual being born with an exceptional ability to register olfactory information at the conscious level — information which is below the level of conscious perception in most people. The conflicts resulting from this heightened awareness are potentially capable of eliciting many of the symptoms of schizophrenia.
3. A chemist in California searched for people with *reduced* ability to perceive specific odors, that is, people with "specific anosmias" to these odors. He went into the business of making and selling test-kits for diagnosing these specific anosmias.
4. An osphresiologist in Canada (me!) noted that the smells to which these people were specifically anosmic are, with one possible exception, related to the sexual function. He asked himself whether specialist receptors may be present in vertebrates as well as the invertebrate insects studied in Germany, and whether a deficiency of any one of them might interfere with the formation of normal social interactions. Note that this is the inverse of the idea picked up from the New York pediatrician.
5. A psychiatrist was willing to consider looking for specific anosmias in his patients with a view to finding a useful diagnostic tool, but he lacked the funds to buy the kits and in any case could divert only a limited amount of time from his day-to-day clinical practice.

Things stopped there.

A comparatively modest sum could have provided the psychiatrist with a research assistant and the necessary test-kits and might well have led to important developments. Evidently, too, with things as they are, the bringing together of such a disparate collection of ideas was largely accidental.

It is unlikely that the man hired by the Foundation as its Itinerant Director would function effectively for more than about 5 years. After that he would be expected to make way for a new man with a similar type of mind but drawn, perhaps, from a different scientific field. The man on retiring would be uniquely qualified for an administrative position in government or industry by virtue of his personal contacts with a very wide range of working scientists, his understanding of how their minds work, and his appreciation of what they hate most in bureaucrats! In a few cases he may become so involved in the field that he will want to develop his own research program. Any university would be fortunate indeed to employ him because his interdisciplinary approach would be so unlike the more narrowly specialized outlook of too many university researchers.

Clearly, the idea of giving an Itinerant Director the individual authority to allocate research funds is unusual. If his terms of reference are to coalesce the activities of geographically and departmentally separated scientists, it is hard to see how else he could function. Given that an even more important objective is for him to mimic the honeybee by cross-pollinating ideas, a few conspicuous successes would be held to outweigh a certain proportion of utter failures. His object is not so much to ensure a smouldering mediocrity as to produce an occasional illuminating blaze.

A 20-year test of this idea would be eminently worthwhile.

INDEX

A

B

C

D

G

H

I

J

K

L

M

P

T